THE SHAKESPEAREAN DRAMATURG

THE SHAKESPEAREAN DRAMATURG

A THEORETICAL AND PRACTICAL GUIDE

By

Andrew James Hartley

THE SHAKESPEAREAN DRAMATURG
© Andrew James Hartley, 2005.

First published in 2005 by
PALGRAVE MACMILLAN™
175 Fifth Avenue, New York, N.Y. 10010 and
Houndmills, Basingstoke, Hampshire, England RG21 6XS
Companies and representatives throughout the world.

PALGRAVE MACMILLAN is the global academic imprint of the Palgrave Macmillan division of St. Martin's Press, LLC and of Palgrave Macmillan Ltd. Macmillan® is a registered trademark in the United States, United Kingdom and other countries. Palgrave is a registered trademark in the European Union and other countries.

ISBN 1–4039–7007–6

Library of Congress Cataloging-in-Publication Data is available from the Library of Congress.

A catalogue record for this book is available from the British Library.

Design by Newgen Imaging Systems (P) Ltd., Chennai, India.

First edition: November 2005

10 9 8 7 6 5 4 3 2 1

Printed in the United States of America.

*To Mum, Dad, Finie, Sebastian, and the
Georgia Shakespeare Festival*

CONTENTS

List of Illustrations   ix
Acknowledgments   xi

Introduction: The Shakespeare Police   1

PART I   PRINCIPLES   9

Section I   The Shakespearean Dramaturg: A Job Description   15

Section II   Dramaturgical Issues: The Theory   29

1   Theatrical Collaboration and the Construction
    of Meaning   31
2   The Text/Performance Relationship   35
3   Adaptation and Authenticity   43
4   "As It Was Originally Done": The Logic Behind
    Historical Reconstruction   46
5   The Nature and Use-Value of History   51
6   Ambiguity and Polyvocality in the Plays   56
7   Authorship, Authority, and Authorization   59
8   Different Languages   65
9   Why Stage Shakespeare?   70

PART II   PRACTICE   73

Section III   Before Rehearsals   75

10   Preliminaries, Casting, and Directorial Vision   76
11   Thinking About Script   88
12   Preparing the Script   95
13   Script-Editing Examples   115

Section IV   During Rehearsals   149

14   Degree and Nature of Involvement   151
15   Tools of the Trade and Research Packets   153
16   Knowing Your Audience: Talking to Directors,
     Talking to Actors   160

17 The Dramaturg in Rehearsal: A Temporal
    Breakdown   165

Section V   Opening and Beyond   181

18 Writing for the Audience (Synopses and
    Program Notes)   183
19 Program Essay Examples   192
20 Talking to the Audience   205
21 The Dramaturg as Advocate   209
22 Evaluating and Sharing your Experience   212

Notes   215
Works Cited   227
Further Reading   230
Index   232

LIST OF ILLUSTRATIONS

10.1 *A Midsummer Night's Dream*. GSF 2000, Dir. Garner:
     Mark Kincaid (Oberon), Saxon Palmer (Puck), Janice
     Akers (Titania), and Thalia Bauldin (First Fairy).
     Photo: Kim Kenney   83

10.2 *The Winter's Tale*. GSF 2001, Dir. Garner:
     Rob Cleveland (Polixenes), Anthony Irons (Florized),
     Allen O'Reilly (Leontes), Tim McDonough (Camillo),
     Janice Akers (Paulina), Carolyn Cook (Hermione), and
     Jessica Andary (Perdita). Photo: Rob Dillard   84

13.1 *The Taming of the Shrew*. GSF 2002, Dir. Garner:
     Gabra Zackman (Kate) and Saxon Palmer (Petruchio).
     Photo: Tom Meyer   120

13.2 *Twelfth Night*. GSF 2000, Dir. Epstein: Bruce Evers
     (Sir Toby Belch), Chris Kayser (Sir Andrew Aguecheek).
     Photo: Georgia Shakespeare Staff   147

15.1 *Comedy of Errors*. GSF 1999, Dir. Epstein: Jonathan
     Davis (Dromio), Peter Ganim (Antipholus), Chris
     Kayser (Dromio), and Linda Stephens (Balthasar).
     Photo: Georgia Shakespeare Staff   155

15.2 *Macbeth*. GSF 2004, Dir. Fracher: Sherman Fracher
     (witch), Marni Penning (Lady Macbeth), Daniel May
     (Macbeth), Alison Hastings (witch), and Bruce Evers
     (witch). Photo: Bill DeLoach   158

17.1 *Julius Caesar*. GSF 2001, Dir. Dillon: Theresa
     DeBerry (Calpurnia), Brik Berkes (Decius Brutus),
     Bruce Evers (Julius Caesar). Photo: Rob Dillard   168

ACKNOWLEDGMENTS

This is a book about collaboration. Not surprisingly, it could not exist without the work of a great many people who have influenced my sense of academia, the theatre, and dramaturgy in particular over the years. I would first like to offer special thanks to those who have helped me work on the manuscript itself: James Bulman, who first said it was worth writing; Jeremy Lopez, who read the book as I was drafting it; Amlin Gray and Cary Mazer, who supplied invaluably detailed notes; Elizabeth Hetzel, who proof-read it; Stacey Colosa Lucas, for help with the pictures. For less direct—but equally valuable—support over the years, I would like to thank James Siemon and William Carroll, who supported my first forays into the staging of Renaissance drama, and to all who worked with me in *Willing Suspension Productions*, particularly Kirk Melnikoff, Mike Walker, Jonathon Mulrooney, Lauren Kehoe, Peter Lurie, Sarah Lyons, Jill Orofino, Michael Arner, and Michael Hamburger.

My sense of dramaturgy in the professional theatre comes largely from my work with the Georgia Shakespeare Festival and I owe a special debt of gratitude to the gifted actors, designers, directors, production staff, and patrons with whom I have been privileged to collaborate in the last eight years, particularly Bruce Evers, Chris Kayser, John Ammerman, Allen O'Reilly, Hudson Adams, Carolyn Cook, Saxon Palmer, Joe Knezevich, Damon Bogges, Margo Kuhne, Kathleen McManus, Janice Akers, Tim McDonough, Courtney Patterson, Rob Cleveland, Park Krausen, Bradley Sherrill, Jonathan Davis, Daniel May, Tara Ochs, Chris Ensweiler, Tim and Kat Conley, Charlie Caldwell, John Dillon, Nancy Keystone, Drew Fracher, and—especially—Sabin Epstein. It has been an honor to work with you all. A special acknowledgment is owed to Richard Garner, GSF's producing artistic director, the source of its extraordinarily familial energy, and a dramaturg's director.

I would also like to thank my former colleagues, staff and students at the University of West Georgia for their support over the years, particularly Kimily Willingham, Jane Hill, Michael and Lisa Crafton, David Newton, Rob Snyder, Randy Hendricks, and Debra MacComb.

Finally, a special thank you to my wife for her support during the writing of this book and—more importantly—during the productions whose rehearsal periods provide the book's impetus and evidence, and to my parents who first instilled in me a love of the theatre.

INTRODUCTION: THE SHAKESPEARE POLICE

It is often observed that the position of the dramaturg is one characterized by its "in-betweenness," defined by what it is not: not playwright, director, or actor, but also not simply literary critic, historian, or theorist. In the case of the Shakespearean dramaturg, the two broad categories between which he or she moves—literary academia and practical theatre—are particularly different and, at least sometimes, proud of and adamant in their difference. Indeed, these are not so much categories as cultures and, like all cultures, they have their attendant ideologies and discursive modes as well as their own values, assumptions, and agendas. Sometimes the differences between literary academia and practical theatre seem so immense that they appear to have little to say to one another, a depressing idea made more so by the fact that it is sometimes articulated by those who have spent the most time attempting to make the two cultures speak to each other.[1] The dramaturg, floating between these two cultural spheres, can sometimes feel like he doesn't quite belong in either, and that neither fully understands or respects his work. This book is, in part, an attempt to address that problem.

What follows aims to go beyond the genre of apology, however, attempting instead to define the job by exploring what it is or can be, and thus moving toward some positive definition of dramaturgy instead of relying on those familiar ideas of placing the dramaturg in the spaces between categories, of defining the role by what it is not.[2] Before I get there, however, I have to acknowledge that some of this liminal sense of the dramaturg's position comes not only from a general uncertainty about the nature of the role, but also from the breadth of backgrounds and training that dramaturgs tend to have. Some, for example, are theatre people trained in college seminars to function as dramaturgs, people who get their feet wet by being attached to actual school productions. Others are literary academics who begin with no experience in the world of practical theatre. These are two broadly different demographics and this book must address both. Since this binary is itself somewhat simplistic, however (many dramaturgs inhabit, not surprisingly, a middle ground, graying still the gray area between the black of the text and the white of the stage lights), it does not make sense to divide the book into sections, some aimed exclusively at literary academics, some aimed at theatre practitioners. Instead I have proceeded as if each section may be of use to all readers, though I realize that some parts will contain more that is "news" than others according to the reader's background, experience, and training. In short, the book inhabits the same liminal middle ground as the dramaturgs whose work it describes, and as such it should provide for a more general readership a case study of the logics and methods involved in bridging the page/stage gap.

Whatever else he or she is, the dramaturg is an intellectual presence in a production (though this is more about perspective than it is raw cleverness), one whose job is to view the show from the standpoint of ideas. It seems especially important, therefore, that I attempt to clarify those ideas that have a direct impact on what a dramaturg does before turning attention to more methodological concerns. Such a rendering seems doubly necessary since the ongoing conversation about dramaturgy seems to draw on broadly dissimilar ideas. There is general consensus on what a dramaturg should do, but a good deal less on how a dramaturg should *think* about what he or she does, and the result is that that general consensus gets quickly muddied when dramaturgs discuss the specifics of their work in the absence of a larger theoretical framework.

To suggest, then, that my goal here is simply descriptive would be disingenuous, since I have a particular agenda that pushes beyond what dramaturgs are and argues for what they should be and why. I want to get at those assumptions about dramaturgy that are held by the dramaturgs themselves and by those who employ them, particularly since it is due to some of these assumptions that some theatre practitioners don't want to deal with dramaturgs at all. Let me illustrate the point with an anecdote. On walking into a rehearsal recently, the director, nodding toward me, said to the actors, "Everybody stop what they're doing. We've been busted. The Shakespeare police have arrived."

It was, fortunately, a joke, but one which contains a problem at the core of dramaturgy or, at very least, at the core of how dramaturgs are perceived, a perception in which the dramaturgs are frequently and problematically complicit. A dramaturg, like a teacher, is *in* authority because he or she is *an* authority, and although dramaturgs are fond of bewailing their lack of power, this sense of authority is still a real force, though not always a constructive one. Indeed, if the dramaturg's aura of authority does not merely result in his or her being shut out of the rehearsal process altogether, that authority can have a crippling effect on the production. This should not be surprising because the areas in which the dramaturg is thought to be an authority— again, particularly where Shakespeare is concerned—are largely untheatrical, inhering instead in matters of author and history, both of which are accessed through a largely monolithic and immutable literary text. In short, the dramaturg is perceived to partake of the authority of this unholy trinity (author, history, and text), and his or her presence in a production is often seen therefore as that of an arbiter of accuracy and correctness. For example, Bert Cardullo, editor of the important essay collection *What Is Dramaturgy?* says,

> a dramaturg is . . . the guardian of the text (presuming there is a text worthy of guarding) as opposed to its "author," a stand in for the playwright. His job is

to know (insofar as this is possible) as much as the playwright—about history, society, culture, and politics, as well as drama—when he set about writing this or that play, and his goal is thus to ensure the theatrical transmission of the playwright's vision, the "making" of the play on stage. (Cardullo, 10)

The general sense of the dramaturg being informed about the history and culture of the moment in which the play was written is important, useful, and unobjectionable, but several other elements loom more problematically in this thumbnail sketch of what dramaturgs are supposed to do. First, there is that sense of the dramaturg as "guardian of the text," an idea so familiar, perhaps, that the eye slides over its implication that the text is somehow under siege, presumably by the director or the actors. In fact, I would say that such a position suggests a deeply problematic privileging of text over theatre, one which assumes that at its best the stage is a kind of conduit through which the text is presented. This kind of assumption is particularly alive in theatrical approaches to Shakespeare both from practitioners and from critics, partly because of the logic inhering in Cardullo's caveat about the text being "worthy of guarding." Few dramaturgs would dispute the Bard's claim to such worthiness.

But who determines what is worthy and according to what standards? More to the point, doesn't this idea that some texts are worth guarding and some aren't suggest two broadly different approaches to the construction of a theatrical product? As a director, I would rather have a text deemed unworthy of guarding, since then I could focus solely on the theatrical product. As a dramaturg attached to a Shakespeare production—and thus to an eminently guardable text—the job is apparently to insulate the text against those theatrical elements that might somehow change it. After all, the text is, Cardullo seems to assume a coded document that, if viewed through the right historical lens, will afford the dramaturg a clear sense of the "playwright's vision," something that can then be transmitted to the stage. Seen from this vantage point, the theatre becomes no more than a kind of receiver broadcasting the author's genius (the transmitter) down the airwaves of the literary text.

In its privileging of author, text, and history over the self-authorizing power of theatre itself, such an approach to dramaturgy, indeed such an approach to how theatre functions, seems to me completely wrong-headed. It is a critical commonplace to say that theatre practitioners are generally wary of theory, and with good reason since theirs is a medium of concrete choices and material conditions, but those concrete choices grow consciously or otherwise from ideas and assumptions that can benefit extraordinarily from a theoretical grounding. How dramaturgs talk about what they do is ultimately of only minor concern to me. More pressing is the sense that flawed discourse reflects flawed thinking and thus flawed practice. How, for example, does one who considers himself a textual guardian bent on transmitting an authorial vision think about and execute the editing of a Shakespearean performance script? Doesn't such an attitude immediately circumscribe the nature and range of any subsequent production, and if that attitude confines the production according to ideas that are theatrically conservative and logically suspect, does it not behoove the dramaturg to try and deconstruct them? In short, shouldn't the theoretically and critically aware dramaturg be fighting *against* this textual-guardian business, rather than for it?

Much of the writing on dramaturgy seems to me on considerably safer ground when dealing with new works, where the cultural status of the text and the author do not loom iconically over the stage as Shakespeare does. Shakespeare is a talisman, a magical presence—or absence—whose place in the theatre is wholly overwritten by his place in every other aspect of culture, particularly his place in education. The result, I think, is a phenomenon in which the theatrical is somehow trumped by the textual and by that larger cultural monolith which is "Shakespeare." The result frequently paralyzes the dramaturg (and ultimately the show itself), reaffirming Cardullo's sense of the dramaturg as authorial and textual guardian, rendering the job of drama-turgy something like running interference for that monolithic and textual "Shakespeare," protecting him/it against the encroachment of theatrical practice. But to go into the rehearsal room with this sense of the job—given the fact that the difference between page and stage is one of kind, not degree, one of *genre*, in which theatrical practice necessarily *constructs* a new artistic product—is to embrace a series of defeats; this dramaturg is able only to give plaintive voice to the blasphemy of what is being visited on the text and on the long dead author. For such a dramaturg, victory is not an option, unless the production were to consist of the cast handing out copies of the play to the audience and telling them to go home and read them. No actual pro-duction can be "straight" enough to somehow emanate solely from the text or the mind of the author, so to the "guardian" dramaturg, all theatre is transgressive, blasphemous, and doomed to failure.

Such a position is clearly not viable. If the dramaturg's authority derives from author, history, and text and is constructed in such a way that his or her function is to preserve them against the onslaught of the stage, the result is not negotiation or collaboration but trial by book. A powerful dramaturg becomes judge, jury, and executioner, the person presiding over whether what goes on in rehearsal is or is not in accord with Shakespeare's "vision" as codified in the text. The less powerful dramaturg—and this is more common—is given no such judicial authority and is instead banished to the program notes, shut out of the production's constructive loop, and wheeled out to do preshow lectures on The Great Chain of Being.

But what if the dramaturg's authority derives not merely from the author/history/text trinity, but stems also from an intellectual investment in the constructive (not merely transmissive or interpretive) energies of theatre itself? Surely, if the dramaturg can manifest a thorough commitment to what theatre is and how it works theoretically and practically, those other elements on which the dramaturg's authority is based (two of them at least, I have lit-tle use for the author himself) might be better used, not as a scale by which to measure the production's *rightness*, but as elemental ingredients that might make for a better theatrical product.

I see such a position not as overturning the conventional investment in a dramaturg's intellectual presence and specialist knowledge so much as fulfilling it. This intellectual presence must be aware of theatre's self-authorizing strategies, its collaborative construction and semiotics, in order to be logically coherent within the medium's operational field. To insist on literary and historical significations in the production of theatre is to misread the genre and render oneself irrelevant, however much one does so with the spurious sense of taking the moral high ground. This book is about applying what we know about how theatre works theoretically to the literary and historical concerns that form the backbone of much of the dramaturg's work, in the hope that a better understanding of the former will make for a more constructive use of the latter. The widespread pursuit of such resultant marriages (the theoretical and the practical, the literary and the theatrical, the historical and the urgent contemporeity of the stage) is designed to—in time—undermine misapprehensions about what we are doing when we stage Shakespeare, and to give a greater use-value to good dramaturgs. In the process, those dramaturgs will slough off the associations of textual guardians and Shakespeare police, becoming less the paralyzing arbiters of accuracy, and more genuine collaborators in an expressly theatrical construction of meaning. In order to effect this, the dramaturg must learn how to best use this embracing of the theatre's self-authenticating nature, finding, for example, new ways to use research tools in support of the production regardless of whether or not such research has any connection to Shakespeare, his culture, or his theatre.

Contrary to popular opinion, then, the dramaturg is not locked in mortal combat with the director. Indeed, I would say to all directors who read this book that the dramaturg is your friend and collaborator, however much that essentially amicable spirit may occasionally be constructively adversarial. The dramaturg's allegiance is to the production, not to the author, the text, or a prior set of literary ideas, and any dramaturg who does not embrace this misrecognizes his or her function utterly. For the director, a dramaturg is a multifaceted resource and an intellectual presence, one to whom the director can turn for help, advice, and opinion on all aspects of how the show is being built or how it is starting to play. The dramaturg/director relationship must be as trusting as that among the director and her actors and designers if it is to be fruitful. Where walls are built between them an atmosphere of hostility and resentment is unavoidably created, resulting at best in missed opportunities, at worst in active subversion of each other and the show. If such a situation arises and the production suffers, one or both parties have lost sight of their loyalty to the company and the theatrical event they are mutually constructing.

Invariably, the barriers between directors and dramaturgs are based on dubious assumptions about where the other person is coming from and what they value. Part of this book's function is to assert the extent to which director and dramaturg—though enjoying different functions and spheres of influence—are on the same page in terms of agenda. The fact that dramaturgs are likely to view the production more intellectually than aesthetically or viscerally (though a truly intellectual approach to how the show constructs meaning will take the aesthetic and emotional factors into account) does not mean that they cannot communicate or that their goals for the show are antithetical. On the contrary, the dramaturg's presence simply allows the director a second set of eyes and ears, which can be trained onto specific aspects of the production, an extra mind to provide thoughtful response to what is or might be going on as the process advances toward opening. Differences of opinion are inevitable, but they should be tackled as is any other disagreement encountered during collaborative work: with discussion, thought, experimentation, and decision. The good dramaturg will always leave the final call to the director, will respect it and work thereafter to make it play. We dramaturgs might not always look or sound like other theatre folk, but we come in peace, and we're here to help.

I begin with a brief outline of what dramaturgs generally do, breaking the job down into its component parts, before moving to a discussion of the theoretical principles that underscore or otherwise involve the dramaturg's work as I envisage it. Since this is a book about praxis—the embodiment of theoretical principles in the material world—I will neither reinvent the theoretical wheel nor overburden the reader with convoluted abstraction. Instead I will discuss in general terms the consequences of certain key theoretical principles and debates for one involved in practical theatre. Much of this first part deals with the problematizing of familiar assumptions about the author/history/ text trinity, and about the implications of such ideas for theatrical production in general and dramaturgical work in particular. The core conclusions of this section—that all performance is adaptation, for example, and that theatre authorizes itself—form the intellectual basis for the second part of the book, which deals with the concrete and specific conditions of the dramaturg's work.

This second part moves along a generic timeline, plotting the way the dramaturg's work shifts in the course of a production, beginning with pragmatic concerns such as contract agreement, preliminary conversations with the director and other members of the production staff, and moving into a detailed consideration of script editing. This section includes a series of short annotated extracts from play scripts, each attempting to manifest in terms of actual choices and decisions the logic and rationale of editing already presented in principle. The following section details the dramaturg's work during the rehearsal process, how the aforementioned issues of his or her own authority are negotiated with actors, and how the job necessitates a withdrawal from a position of relatively high collaborative visibility as the production progresses toward opening. As the dramaturg moves from constructor to audience at the tail end of this stage, so the dramaturg's role changes again once the show has opened. The final section thus charts the dramaturg's options in matters of those more peripheral dramaturgical obligations such as the writing of program notes, participation in audience lectures and talkbacks, and the sharing of one's ideas and experiences with other dramaturgs.

Everything in this second part is premised on the more theoretical discussion that precedes it and the two parts are thus less distinct than they may initially seem, each being shot through with the concerns, assumptions, and demands of the other. The overarching concern of the book is to push the dramaturg out of the rehearsal room corner, to shed some of that self-consciously marginal status by affirming what I believe the job is, rather than defining it by what it isn't. For the dramaturgical wall flowers to take a real part in the theatrical dance, however, they must also shed that aura of being the police-men of the text, the lawyers representing Mr. Shakespeare come to do their damndest to shut the production down if it doesn't get it "right." As I hope to demonstrate, my hostility to such regulatory notions of dramaturgy are rooted not simply in the collaborative spirit of theatre, but in a theorized notion of what texts, authors, and assumptions about history are. If dra-maturgs are uniquely positioned to scrutinize a production for its intellectual elements and consequences, then they must begin by applying that same intellectual rigor to the nature and assumptions of dramaturgy itself.

I should add that my aversion to that restrictively textual version of the dramaturg's authority is not just about producing good theatre. It's also about a version of the world, and the place of theatrical art within it, because the ubiquitous textuality of Shakespeare, that iconic sense of "the Bard" enshrined in schools and colleges, is invariably hierarchical and exclusive. That sense of "correctness" then, which some dramaturgs see themselves policing, aligns them and their product with a cultural elite, with the implicit assumption that "real" Shakespeare is reserved for those who deserve it. From such a viewpoint, productions which don't manifest Shakespeare that is (somehow) textually or authorially "pure" (and I use that loaded word advisedly), are not merely suspect; they pander to those unworthy of the "real thing," trivializing it, even corrupting it. But such a position is distasteful and intellectually untenable. If, instead, we can escape the notion that the dra-maturg's job is to somehow ensure the transmission of a textually authoritative Shakespeare, then perhaps we can rediscover the plays not as the trappings of cultural nostalgia and elitism, but as cultural products that communicate—and therefore entertain, move, provoke, instruct, and so forth—right NOW and for whomever comes to watch them on stage.

Theatre is a uniquely local phenomenon, rooted in actual places, commu-nities, and moments of time, in ways other artistic media generally aren't. Any discussion of dramaturgy, then, must be up front about the specifically local practices that determine what the job is or can be, since it varies from show to show and company to company, according to factors as diverse as individual personalities, local demographics, economic and administrative structures within the theatre, and so on. This book makes no claims to uni-versality or even to a general ideal. Instead, it draws on a particular model—that of the Georgia Shakespeare Festival where I have worked for eight years and am now resident dramaturg—in order to make specific what can easily dissolve into abstraction and, more importantly, to insist upon the essentially local nature of all theatre, especially regional theatre, which is largely attended by the community of people who live close by. This specially anchored sense of itself makes the regional theatre or Shakespeare Festival a particularly fruitful workplace for a dramaturg, since the essentially collaborative nature of dramaturgy (mediating between literary critics and actors or among historians, theorists, and directors) thus extends not just to the company members, but to the extra-theatrical community: the local press, the business world, the schools, the other theatres, the audience itself in all its diversity of background and opinion.

As the dramaturg is a resource and an intellectual presence, so is he or she of necessity a teacher. By this I don't simply mean that the dramaturg is there simply to explain primogeniture to actors or to hold forth on the politics of the Elizabethan monarchical succession in pre-show lectures. I mean that the dramaturg's educational function should be about opening up theatre to new creative energies, about making the stage a place for the development and exchange of ideas which reach beyond that stage (and beyond Shakespeare) and touch the hearts and minds of the community at large. The theoretically informed and theatrically constructive dramaturg can then aid not just the production, but the community, in its intellectual deconstruction of old prejudices (aesthetic, cultural, political, and so on) through the establishment of a more approachable, inclusive, and vibrant Shakespeare, a *theatrical* Shakespeare. If something of that textual, monolithic Shakespeare—and the dramaturgical policeman bent on protecting him—has to be lost in the process, so be it.

PART I

PRINCIPLES

The purpose of this first part is to establish a theoretical framework for what follows by providing a general overview of some of the driving issues and concerns that affect the place of the dramaturg and the nature of his or her job. Since the readership of this book is likely to come from different backgrounds, I have attempted to keep the use of highly specialized or professional language (what is sometimes derisively dismissed as mere 'jargon') to a minimum, trying instead to discuss the assumptions at the heart of theoretical analysis in a way that keeps them anchored in those practical conditions of the theatre that will dominate the latter half of the book. It is possible that some of the ideas presented here may consequently suffer a little from over-simplification, though not in ways that significantly affect their overall content or the way they will function as guiding principles for the study of practical Shakespearean dramaturgy that follows.

As I said in the introduction, my experience as a dramaturg has been dominated by a single company, the Georgia Shakespeare Festival, and it seems fitting to say something about that company as a preface to my discussion of the job itself, since that job is, like all theatre, defined by the material and cultural conditions of the company, the audience, and the surrounding community.

GSF was founded in 1986, and though it initially played in a tent erected seasonally, it moved into its permanent home on the Oglethorpe University campus in Atlanta in 1997. The current building (which also contains offices and rehearsal rooms) has a 509-seat theatre to which GSF has full access six months out of the year. It has an annual budget in excess of 1.2 million dollars and is rated a League of Resident Theatres (LORT) D theatre, second in theatrically rich Atlanta only to The Alliance. The company stages a three show summer season in rotating repertory, and a fall show, as well as staged readings and symposia tied to events on the main stage. In 2004, GSF revived a former production to be staged in Atlanta's Piedmont Park, and there are plans to create a second space to which the company will have access year round in the near future. Annual audiences total approximately 60,000 people, and special education programs, including touring shows, reach about 100,000 students per year in neighboring schools and colleges. That said, the attention the festival receives in the form of reviews is largely local, and though it is one of the largest theatrical operations in the southeast, it does not draw much of its audience from outside the state, compared to, say, the Alabama Shakespeare Festival located in Montgomery.

GSF's acting company is all professional, a little over half of the cast of any given show is in Actor's Equity, and some who have worked with the company for many years are classed as associate artists. Most of the actors are based locally though some are brought in from New York and elsewhere. Directors are brought in from all over the country, though many return to the festival in subsequent years. The producing artistic director and cofounder, Richard Garner, was trained as an actor at San Francisco's American Conservatory Theatre, studied theatre management in New York, and recently served as president of the Shakespeare Theatre Association of America (STAA).

GSF's use of dramaturgs has become more formalized over the past decade, though they are still hired on a per show basis (i.e., as freelance production dramaturgs rather than staff positions). A letter of agreement is issued to the dramaturg, specifying basic responsibilities but making no stipulation about the percentage of rehearsals to be attended, or requiring the generation of a production file or book (something often identified specifically in contracts made for dramaturgs who serve as company staff). Payment is a respectable honorarium, which has climbed steadily over the last five years; however, it remains significantly less than the sum paid to directors and designers partly because of the relatively few demands made of the dramaturg in the contract. Having worked on a dozen mainstage shows and other projects, I became the company's first resident dramaturg in 2004, a position that gives me greater involvement in company-wide issues such as season planning.

Understanding GSF and its use of dramaturgs requires consideration of The Atlanta Shakespeare Company (ASC), which performs at the New American Shakespeare Tavern, an indoor, quasi-Elizabethan space, elsewhere in the city. The ASC is a smaller outfit in terms of payroll and audiences, though they have complete control of and access to their performance space. The ASC (or the Tavern, as they are known locally) does not employ dramaturgs, something bound in part to their basic philosophy and approach, since the Tavern is an "Original Practices Playhouse." As they say on their website:

> On a practical level, ASC's philosophy is guided by an unshakable belief in the playwright's original voice. Consequently, decisions about costumes, props, sets, food, drink, tables, chairs—virtually every conceivable manifestation of an ASC production—have been guided by a single overriding voice for almost fourteen years: the voice of the playwright William Shakespeare. This commitment to a core aesthetic centered on the playwright's original intent is in clear contrast to the majority of America's other classical repertory companies where texts are routinely altered in favor of an "updated" or "conceptual" approach. (http://www.shakespearetavern.com/MoreAboutUs.html)

I cite this description of Atlanta's "Other Place," because it seems to me that the two companies in part define themselves against each other, and though this is generally achieved without rancor, they do manifest significantly different approaches to the staging of Shakespeare. The Tavern does not use dramaturgs except insofar as members of the company provide historical research into the understanding and performance of the script in an expressly Renaissance context. My sense of the essentially local and material nature of dramaturgy, then, grows out of this juxtaposition of the two theatres, one of which (GSF) embraces a progressive and conceptual approach to productions—and an attendant notion of dramaturgy; the other (The Tavern) claims a historicized approach grounded in the First Folio, manifested by historically appropriate live music, Medieval and Renaissance costume, a unit set stage, broadly Elizabethan acting techniques—and a historically reconstructive notion of dramaturgy.[1]

How companies use dramaturgs, thus, has significant implications for their sense of what they are doing when they stage Shakespeare. Indeed, despite the rhetoric of the disinterested dramaturg, the job is *fundamentally ideological*, and the claim not to have an ideology (like the claim not to have a "concept" for a show) is—consciously or otherwise—a ruse, and a thin one at that. Dramaturgs go into a show with a set of assumptions, principles, or beliefs, which translate into the practical work of the production, research, script editing, and so on. Some of these assumptions are more intellectually coherent than others, and it is thus vital that we identify and interrogate our guiding principles since they shape the kind of work the dramaturg does, and the kind of production that results.

As a rule, smaller companies, particularly those that, like the Tavern, gravitate toward what they consider to be historically reconstructive productions, tend to use dramaturgs sparingly, if at all, something that probably has as much to do with budget as philosophy. Locally, for instance, the North Carolina Shakespeare Festival does not use a dramaturg, and though the Nashville Shakespeare Festival does they are currently unpaid. Compared with midsized and larger outfits dealing extensively or predominantly with Shakespeare, however, GSF's use of dramaturgs seems fairly typical, always granting the particular demands and conditions of the larger community and attendant budget issues. The Alabama Shakespeare Festival, for example, has a part-time, salaried, season-wide dramaturg, Susan Willis, whose experience as a production dramaturg, though it varies from show to show and director to director, is broadly like mine. Some companies, such as the Colorado Shakespeare Festival, which are resident on university campuses, pay dramaturgs drawn from the faculty and staff of their host institutions. The Orlando-UCF Shakespeare Festival uses production dramaturgs and pays them a small stipend, while the Seattle Shakespeare Festival uses a single dramaturg for the entire season but pays her on a per-show basis. The Utah Shakespeare Festival uses a season-wide dramaturg (Michael Flachmann) recruited from literary academia, who is salaried annually but retains an academic position elsewhere. The Oregon Shakespeare Festival (Ashland) employs both season-wide dramaturgs (who work on three to five productions) and production dramaturgs who work on a single show. Other large

American companies such as the Guthrie, which produce Shakespeare as part of a broader theatrical mission, employ several dramaturgs as staff working under a literary manager, with production dramaturgs also hired on an ad hoc basis to work on individual shows.

Some theatres give slightly different titles to the dramaturg, or include dramaturgical responsibilities under related job descriptions. Washington D.C.'s Shakespeare Theatre, for example, employs a literary manager who "provides dramaturgical support to directors and the entire artistic team for the five-play season and the ReDiscovery Series. This support may include historical, critical, and literary research, pre-production conversations with the artistic team, textual/editorial assistance to directors and authors, script preparation, publication articles, and audience enrichment."[2] The Stratford Festival in Ontario, Canada, uses the word "dramaturgy" to describe work done in the development of new plays, where as the person who performs the dramaturgical work on Shakespeare is titled the assistant director (A.D.) and works on a per-show basis. The blurring of the line between dramaturg and A.D. is not uncommon—particularly where dramaturgy is more linked to new material than familiar plays—though the difference in name generally has few implications for the nature of the job, and what I say about dramaturgs working on Shakespeare seems largely relevant to what the Stratford Festival calls the A.D. Some major companies such as The Chicago Shakespeare Theatre do not use dramaturgs but expect other members of the company such as the text coach to cover at least some aspects of the work I assume to fall under the dramaturg's purview. Massachusetts's Shakespeare and Company do not currently use dramaturgs for budgetary reasons, though they say they would like to, but they get much of what I would call dramaturgical work from Tina Packer (the company's founder and artistic director) and Dennis Krausnick (the director of training), while also maintaining contact with academic Shakespeareans such as Frank Hildy, Andrew Gurr, and Stephen Greenblatt.

In the United Kingdom, where the word "dramaturg" is less commonly used (and when it is, it is often spelled and pronounced "dramaturge"), the question of whether or not a dramaturg is employed is often similarly bound to issues of company and budget size. Where dramaturgs are used, the job often falls under the jurisdiction of a Literary Department, as is the case at The National Theatre. Here the work—which includes commissioning new plays and dealing with playwrights as with the season-wide dramaturgs at major American theatres—also involves dealing with the play as text, and all attendant research and scholarship in matters of history. The Royal Shakespeare Company has had a Dramaturgy Department but has, apparently, found the difficulties of being in rehearsal while simultaneously fulfilling more managerial responsibilities too taxing for their current staff structure. As I write this, the company seems to be moving toward what would be considered a Literary Department, and while production dramaturgs will continue to be used, they are likely to be hired on an ad hoc basis working, effectively, as freelancers.

There is certainly a sense in some theatres (particularly in small theatres) that a dramaturg is a luxury item in ways a set designer, say, is not. This assumption is understandably tied to the measurable visibility of the work a set designer does in the construction of the theatrical product, as opposed to the work of the dramaturg that stays, for the most part, under the audience's radar except when it appears in the form of lectures and program notes. Part of my purpose in this book is to query this dramaturg-as-luxury-item assumption, but it is also—and more insistently—to counter certain notions of dramaturgy that have led to misconceptions about the role, misconceptions that are sometimes perpetuated by the dramaturgs themselves.

My anecdotal sense of companies that don't use dramaturgs where the issue is not obviously one of budget suggests that their decision is often driven by a faith in the transparency of the text and its playability. This can work in two paradoxical directions, since this textual transparency is claimed both by companies who insist on a historicist approach to the plays (and quasi-Renaissance productions) and by those who repudiate history entirely in favor of seeing the plays in terms of a transhistoricist contemporaneity. The former dismisses what it considers "concept Shakespeare," the latter "museum Shakespeare," and both see the dramaturg as somehow attached to both flawed approaches. I think both of those positions are reductive and smack of anti-intellectualism, not least in their targeting of dramaturgs as somehow emblematic of other companies' dubious notions of what it is to stage Shakespeare, but I concede that dramaturgs often feed such misconceptions directly. Often that sense of the dramaturg as Shakespeare police, a presence whose raison d'etre is authoritarian and stiflingly antitheatrical (whichever end of the spectrum the company's notion of 'theatrical' inhabits), originates in stereotypical assumptions about what academics seem to want from Shakespeare on stage: usually an attempt to somehow reify the historical and textual original. Though this version of dramaturgy is something of a straw man for theatre practitioners intent on validating their own process and methods (and budget), it does sometimes get manifested by actual dramaturgs. This book argues against such a version of dramaturgy explicitly, hopefully playing its part in encouraging some companies to reconsider the nature and use-value of a good dramaturg.

What is most telling about what I have learned from the companies I have spoken to who *do* use dramaturgs on a regular basis is that the vast majority of them cite the same basic set of requirements and expectations of the people they hire to fulfill the job. While different companies will emphasize different aspects of the dramaturg's role, the vast majority is clearly working from a model very close to that pursued at GSF and used as the core of this study. Where the dramaturgical work is subsumed under a different title such as assistant director or literary manager, even where no individual is hired to function as a dramaturg *per se*, most theatres still expect the work that I call here *dramaturgy* to be performed by someone, or by the shared labor of various people within the company, be they producers, directors, actors, or voice/text coaches. In such cases, the "dramaturg" I address here may actually be a team, possibly even a large one, and my analysis and advice should be modulated accordingly, though most of it will remain, I believe, broadly relevant and useful.

Since this is the first extended study of the Shakespearean dramaturg's role that I know of, it is necessarily tentative and exploratory rather than expository or didactic. Initially I thought this might be a problem, but I have come to think that this more investigative approach to the material, its tendency to say "here's what I've done and why: see what you think," is fitting to the nature of the job or function under consideration. Publication is, after all, a kind of performance and, like all performance, this book can be in no way definitive of the work of all dramaturgs in all places now, let alone in the future. The role of the dramaturg is rooted in material theatre and the world of ideas, both of which are constantly evolving, so the job itself must be in a continual state of flux as it adjusts to the needs of the historical, geographical, and cultural communities in which it is situated. At various points in the book I have suggested that the dramaturg is an embodiment of theatre's collaborative nature, and I have already observed in my introductory acknowledgments the extent to which this study is itself the result of collaborative work over the years. But there is more to it than that, since the book itself is not an end but a part of an ongoing process, like the editing of a script that is then sent on to a director for notes a month before the first scheduled read by the cast. I have tried to answer certain questions, but if all I have actually achieved is to put the questions themselves into broader circulation, that is OK too. I would like to think, after all, that the book will continue to generate further collaboration (between dramaturgs, among actors, directors, scholars, audiences, and the like) as its methods and assumptions are tested and interrogated in both theory and practice, generating in the process a more intellectually cogent and more practically constructive dramaturgy.

SECTION I

THE SHAKESPEAREAN
DRAMATURG: A JOB DESCRIPTION

It has been said that good directors don't need a dramaturg. Part of
the agenda of this book is to demonstrate why such an assumption is wrong,
but to tackle the question properly we must first decide what a dramaturg
is. The roots of modern dramaturgy are largely twentieth century and
German, but the jobs done by the dramaturg are considerably older and of
more eclectic beginnings.[1] Much that today falls under the purview of the
dramaturg was once the domain of the actor/manager in the eighteenth and
nineteenth centuries, and a case can be made that even as far back as Medieval
and ancient Europe there were people who functioned in ways aligning
themselves with the modern dramaturg. Most of these jobs might be considered managerial, though I mean that in an artistic and critical sense, not
merely organizational or financial. The German model that has begun to
establish itself in the United States embodies much of this managerial spirit,
and some dramaturgs have become the heads of prominent companies as a
result. These dramaturgs focus not just on individual productions, or even
seasons, but also on the long-term aesthetic and intellectual purpose of the
company. As such these dramaturgs are crucially involved in the commissioning
and developing of new plays, and in the large-scale building of an artistically
invested community.

Such goals, though admirable, are outside the focus of this study. I am
concerned here with the work of the production dramaturg—someone who
works on a single show—and with the specific demands of staging a Shakespeare
play. As with company or season-wide dramaturgs and literary managers,
however, production dramaturgs are similarly anchored in the geographical
and cultural specifics of the company at large, and something of the company
dramaturg's larger dramaturgical vision informs their work in ways worth
touching on, albeit briefly.

Not all theatre people agree on the precise nature of the dramaturg's role,
and the production dramaturg may consequently find himself shut out of
some aspects of the work (script editing, for example) in which he thinks
he should be involved, though my own informal survey of companies that
use dramaturgs suggests a good deal more common ground than difference.

As a rule, my impulse is to press all prospective dramaturgs to embrace as many of the job's aspects as the situation will permit, always accepting that specific company precedents or directorial assumptions may place limits on the dramaturg's range. What I offer here is a fairly inclusive representation of the job, one that fits the way most companies use the position, and one from which I think companies (or directors) with a more limited conception of the role can benefit. Indeed, this book might be used as a way for dramaturgs in more constrained circumstances to press for more responsibility. For further reading on dramaturgy in general, the reader might consider some of the excellent recent publications on the subject, particularly *Dramaturgy in American Theater: A Source Book* (edited by Jonas, Proehl, and Lipu), the recent special issue of *Theatre Topics* on dramaturgy (13.1, 2003), *What is Dramaturgy* (edited by Bert Cardullo), and the other material listed in the section Further Reading.

One of the best known dramaturgs of the twentieth century was the playwright Bertolt Brecht: as a man, as a writer, as a theorist, political activist, and theatrical innovator, Brecht is an excellent model for anyone who wishes to be a dramaturg. The dramaturg is scholar and theatre practitioner, a historian, a thinker, and an artist, someone invested both in the material conditions of the stage and in its intellectual implications. This is one of the reasons that many successful dramaturgs are also playwrights, people who are creative but whose prime medium is a specifically theatrical language. Like the playwright, the dramaturg—particularly the Shakespearean dramaturg—is a poet, sensitive to the functions of language in all its aesthetic and emotive power. But playwrights also have to function as scholars. They have to be grounded in their discipline, knowing its history, being aware of what has been written before and how their ideas have been approached on stage in different periods. They have to be structurally oriented, seeing patterns and arcs in the play (the play's *dramaturgy* in the sense of how the play functions as drama), reading the text as both actor and audience in order to see how it works, what makes it move and, when it doesn't work, how to fix it. In short they have to be able to see a play as an intellectual and emotional—even a spiritual—property. They can thus evaluate its purpose, method, and value for the community who will see it, and for the company that will absorb the play into its larger sense of identity. All this makes for an excellent dramaturg.

The dramaturg's function where new plays are concerned, particularly if they are still being developed, can be inferred from the previous paragraphs, but dramaturgs are often called upon to work on older plays and their work in such cases is markedly different. This book focuses on the work of the Shakespearean dramaturg, though Shakespeare is really a name on which to pin a certain kind of drama, one that is familiar to both literary scholars and theatre practitioners because of its immense status as a cultural icon. Much of what I say in this study, while using the specifics of Shakespeare's plays to ground my arguments, might equally be applied not only to the dramaturgy of other Renaissance writers like Marlowe or Jonson, but to that of markedly different periods and cultures from the Medieval through the Restoration, the eighteenth and nineteenth centuries, to the present. Of course, the nature of the dramaturg's work will shift as the language and historical context of the play alters, but many of the precepts I underscore here are equally applicable to more modern works and, if the issues of language and culture are adequately weighed, may even be applied to the staging of plays from more radically different periods, places, and languages. My use of Shakespeare as a model is due to the prevalence of his work on the modern stage and because he provides an unusually rich array of dramaturgical issues, not a claim that the notion of dramaturgy I am to develop applies only to him.

Having said that, it is Shakespeare's overwhelming cultural standing and the primacy of a historically defined language that makes the role of the *Shakespearean* dramaturg unique and invaluable. Present day actors and directors are committed to the present, as they should be; theatre is about communication in the performative moment, to an audience of the performers' contemporaries who necessarily struggle (consciously or otherwise) to connect with plays that are wholly different in form and method from drama that is written today. Shakespeare's plays are built around different notions of acting, and comparably different notions and technologies of theatre. The plays are steeped in ideas, logics, and belief systems that are foreign to the present, and are saturated by topical debates and observations that, if we recognize them at all, tend to reinforce the Otherness of the plays and the period that produced them. Above all, these differences of theatre and culture are enshrined in every word of the plays' dialogue. It is not merely context that makes Shakespeare strange, it is utterance, and since utterance is one of the most basic building blocks of theatre, this poses a problem. If not tackled directly, the theatre that follows from this problematically arcane language is likely to be dead, impenetrable or—worse—that thoughtless "reconstruction" of the play as a kind of romanticized snapshot of the past ("museum Shakespeare"), an exercise in cultural nostalgia that is both dead *and* impenetrable.

Even if we are focusing only on the production dramaturg's work and not on the larger concerns of repertory development and new play commission, it is thus vital to recognize that the dramaturg—like the actors and director—is invested in the NOW of the theatrical moment, and is thus working for the present, living audience who will attend the show. The dramaturg's paramount concern, therefore, is making the play work in the present, for the living, rather than being interested solely in the archaeology of the past or in the ways that performance can be considered an exploration of textual or theoretical ideas. When staging Shakespeare, then, the dramaturg has to work to find what makes the play immediate and pressing, the playwright's cultural status and the familiarity of the material being inadequate reasons to stage the play for the community at large. Part of the dramaturg's job—contrary to the ways the role is sometimes viewed as bound to the archival and the academic—is to always be alert to that theatrical NOW, urging the play to speak to the present, however old and familiar it may be. Good Shakespeare should always feel new, refreshed, and it is part of the dramaturg's responsibility to point out those instances when it is not.

One alternative to slavishly and unknowingly "reproducing" Shakespeare as it was originally done (something that is ultimately impossible, even if it were desirable) is to completely rewrite the play in a modern idiom. There is absolutely nothing wrong with such an approach, and I will argue that certain liberties have to be taken with the text to make it playable and to open up opportunities for discovery and invention. But Shakespeare is crucially *about* language. To merely use a Shakespearean plot (something Shakespeare usually borrowed from other sources himself) is only obliquely to do Shakespeare. Shakespeare's power, complexity, ambiguity, and beauty, his exploration of political positions, his study of selfhood and desire, his expressions of pain and loss, of laughter and community, his contested and contesting utterances on every subject ever attributed to him are all manifested by, enshrined in, and inseparable from the words of the plays. While I freely admit that language is only one component of what makes a theatrical production, that language is essential—not uncut or unedited, but still essential—to the staging of Shakespeare at least in the general cultural sense of what *Shakespeare* is. It is matters of language, and of understanding that language's original resonances (even if, as is perfectly reasonable, those resonances will be reworked by the production) that the dramaturg conventionally is seen to be essential.

That said, however, the dramaturg is an expressly theatrical collaborator who is committed to the theatre's self-authorizing and communicative dimension and she cannot thus be grounded only in text, in history, or in the author. As the production is neither simply a window on the past nor an homage to the playwright, so must the dramaturg be a facilitator of those theatrical energies and technologies that go toward the construction of the production as an expressly local and contemporary experience.

Before I talk about the specifics of what the dramaturg actually does, let me first articulate some of my own assumptions about the nature of theatre, which are crucial to my sense of the dramaturgical function. I go into greater detail in explicating and justifying these assumptions in the second section ("Dramaturgical Issues: The Theory"), but for now let me state them simply in axiomatic form:

1. The dramaturg owes his or her loyalty not to the author or the text, but to the production, the piece of theatre that he or she has a hand in creating.
2. The dramaturg is committed to making the production function effectively as communication in the present rather than being invested solely in an investigation of the past.
3. The performance does not "realize," "complete," or otherwise "fulfill" the script, the difference between stage and page being one of kind, not degree. All performance of a known play is, therefore, also adaptation, and there can be no "straight" staging of any given text.
4. The performance is self-authorizing. In other words, the theatrical event authenticates itself, and is not merely a reflection of some prior work (a text, for example, or a set of ideas about that text) to which it can be

compared and deemed inadequate or unfaithful. Performance is constructive as well as interpretive, so what appears on stage must always be thought of as a *response* to a text, not a *transmission* of that text.

5. Performance is necessarily collaborative and this vital collaboration in part defines the artistic nature of theatre. As such, the production cannot be dominated by text or author, and must be seen as specific to—and constructed by—the culture, taste, community, and locale in which the show is staged.

With these few principles in mind we can begin a consideration of what is expected of the dramaturg as a theatrical collaborator, and what skills he or she should bring to the rehearsal room that will help to inform and build the theatrical event. In this section, I break down the Shakespearean dramaturg's job, before returning to the question with which I began this section: why does a good director need one?

Among the dramaturg's greatest assets is his flexibility, not only with regard to the generally collaborative nature of theatre, but in the very nature of what constitutes his job. There are as many types of dramaturg as there are directors, probably as many as there are productions, the job being significantly inflected by the conditions of the company and the personalities of its members. What follows is thus a provisional list of what is often expected of the dramaturg, though the degree of emphasis each element receives will vary from show to show. It is generally in the dramaturg's best interest—and consequently in the best interest of the show—if he or she can be proficient in all the following areas. By embracing all these aspects of the role, dramaturgs can claim a genuine and extensive involvement in the show and in the business of theatre generally.

Textual Specialist

One of the first duties of the Shakespearean dramaturg concerns matters of text. This is not to say that the dramaturg's job is to "protect" the text from the director or from elements of the show that seem overly adaptive. I address the question of the dramaturg's attitude to the text later, but for now it is sufficient to say that the dramaturg needs to know the text on which the production is based inside and out. As well as being familiar with matters of plot, characterization, imagery, and so forth, the dramaturg should also know (or have easy access to) key information about textual variants, contested line ascription, editorial debate, and so forth. Much of this information is readily available in the introductions and textual notes to good editions of the plays, and the dramaturg should have several on which to draw, since they vary in their editing practices. Much of this information may seem too minute or esoteric for use in actual production, but the words of the script are the foundation on which the other elements of the production tend to be built, and the more the dramaturg knows about these words, the better. Some plays, moreover, such as *Hamlet* or *King Lear*, exist in radically different early texts, and are usually published in ways combining elements from each version. The dramaturg will often find it useful to consider alternate versions (including early editions from Shakespeare's own lifetime, which are sometimes dismissed as "bad" texts by scholars), when dealing with issues as different as preparing the initial script, clarifying lines for actors, cutting to decrease running time, or helping designers find key images that will help to shape the set or costumes.

VERBAL SPECIALIST

It is the dramaturg to whom the director and actors will turn for clarification about the meaning of words and phrases in the script. Beyond the larger textual concerns mentioned above, the dramaturg needs to be steeped in the significance, nuance, and feel of language itself. Footnotes in editions of the play sometimes fail to clarify what key words mean, or do so inadequately. Being familiar with these words and having access to good dictionaries can be invaluable to actors, particularly if the dramaturg can offer meanings for a word or phrase other than that which is obvious or most common in the present. Beyond issues of semantics, the Shakespearean dramaturg needs to possess a verbal sensibility, an awareness of how words sound and their attendant aesthetic values. The dramaturg needs an ear for rhyme and meter, even though the speaking of the lines will largely fall to the actors, director, and (if one is attached to the show) the voice coach. The aesthetics of Shakespeare's language are tied both to his cultural status and to the actual meaning of the lines, and thus loom large as points of dramaturgical focus.

RENAISSANCE HISTORIAN

Regardless of where or when the production is to be set, it is always useful to have insight based on the culture and performance conditions in which the play was first written and staged. The show need not attempt to recreate such culture and performance conditions (and in real terms it can't) for awareness of them to be useful. As with much of the dramaturg's job, knowledge about Elizabethan and Jacobean culture (anything from matters of religion and law down to details about sanitation and diet) or playing conditions (acting styles, audience composition, stage and theatre layout, and the like) all help to give the actors, director, and designers *options* by enriching their sense of how such things may have inflected the way the play's meaning was originally constructed.

THEORIST

Unless the framing concept of a given show demands that such issues be foregrounded in the rehearsal process, it is unlikely that the dramaturg will be required to talk about literary and performance theory during the preparation of a show. Having a grasp on theory, however, helps to create a richer intellectual awareness of what a show is—or might be perceived to be—doing.

A better understanding of the relationship between text and performance in the abstract, for example, will help the dramaturg navigate the contested issues of authenticity and authority as she is editing the performance script. Similarly, a sense of the politics of gender and race on stage, though rarely something the dramaturg will hold forth on at length, will help when thinking through very specific issues such as cross-casting, politically charged matters such as anti-Semiticism in *The Merchant of Venice*, or the representation of female sexuality and independence in *All's Well That Ends Well*. The academic discourse of theory may have to be kept fairly tightly leashed in the rehearsal room, but it will help to organize the dramaturg's thoughts on important matters both political and aesthetic. It is also a valuable tool in pre- or postshow lectures and audience talkbacks.

Above all, as its primacy in this book suggests, a grasp of relevant theory is an essential component of the dramaturg's collaborative and educational arsenal, since it demands an intellectually coherent sense of the job, the way theatre constructs meaning, and the way that the two can work together in the interests of better productions.

Production Historian

As well as having a firm footing in the Renaissance, the Shakespearean dramaturg needs to have access to the production history of the play itself. There is a sense in which a production is in dialogue with all previous productions, and while directors and designers generally do not like to imitate other people's work directly, a knowledge of how the show has been approached before is often very useful, if only to provide reasons to do things differently. This is particularly true of well-known cruxes in a play: how does one stage the bear in *The Winter's Tale*, the precipitous marriage of Isabella to the Duke in *Measure for Measure*, or whether Hamlet knows his "Get thee to a nunnery" speech to Ophelia is being overheard by her father and the King? Again, good editions of the play generally include some stage history, other books are devoted solely to the plays in performance, and many periodicals contain extensive reviews of Shakespearean productions on stage and film.[2] It can be very suggestive to discover how pre-twentieth-century productions approached a play, so the dramaturg should not limit himself to consideration of the most recent offerings by the Royal Shakespeare Company (RSC). By the same token, even minor productions by community theatres or students can contain striking or innovative elements the dramaturg might want to bring to the director's attention.

Literary Critic

A tremendous amount of critical writing on Shakespeare and his contemporaries is generated every year in the form of books and journal articles. Some of this output will be of little interest to the dramaturg because it will be of little interest to the actors and director, but performance is a form of publication, and as all performances are in dialogue with each other, so are all performances in dialogue with textual criticism. Sometimes the gap between the two is so vast that they seem to be (and in real terms are) speaking different languages, but that is not to say that critical ideas rooted in the play as text cannot be of value in shaping the play as performance. Moreover, the popular assumption that literary criticism is all about unplayable formal analysis is wrong, much of today's criticism being vitally engaged in pressing cultural and political ideas that may prove invaluable when shaping a production's approach.

Again, proper use of this material (appropriate sifting of the studies and articulating their ideas or observations in ways theatre practitioners can use) gives the production a wider range of options and a greater degree of self-awareness. Use of literary criticism can help to build a crucially intellectual dimension to a production—something that can be easily sidelined in the contemporary rehearsal hall where emotion is often assumed to be paramount—and serves as a way in which the dramaturg mediates between (or collaborates with) academic discourse and theatrical practice.

General Researcher

Sometimes issues arise that demand knowledge from outside both the play and the Renaissance. The dramaturg can serve in the capacity of a more generalized researcher when, for example, actors in a production of *Richard III* set during the Second World War want to know how German army officers saluted each other, and whether such salutes altered according to rank. Such details help to give a production a sense of specificity, but finding out about them may not fall clearly in anyone else's job description. Since the dramaturg serves the production, not the author's text or period, and since the production derives its authenticity from its expressly theatrical dimension, it is appropriate that the dramaturg should extend his research beyond the Renaissance or matters of literary criticism into any area that will help to ground or otherwise support the show.

Sounding Board, Audience, and Intellectual Presence

The dramaturg is uniquely positioned to be an extra set of eyes and ears on the production, and during the rehearsal process—when most of the production and design team have specific areas that require their attention—a good dramaturg can help the director to think through all aspects of the show, including those that stray from matters of text, scholarship, and other more automatically dramaturgical concerns. As opening approaches and the director is subsumed by the various (largely technical) details of polishing the production, the dramaturg—whose role tends to shift as the rehearsal process progresses—can address more general semiotic concerns: is the storytelling clear, is the music too loud or the lighting too dark? Such issues are crucial to the way the show produces meaning, and thus are part of the dramaturg's sphere. While not an authority on the technical aspect of these matters, the dramaturg should have demonstrated by this point in the process that his or her allegiance is to the overall quality of the show and will thus be trusted as an informed audience member who can anticipate how the house (and critics) might respond to the production once it has opened. Furthermore, since actors often approach their characters and their relationships in expressly emotional terms and the director has to work with them accordingly while also keeping an eye on the stage pictures and other aspects of the production, it often falls to the dramaturg to represent a general intellectual presence, someone dealing with the production in terms of the ideas it engages or may produce.

A THEATRE PRACTITIONER

As Jane Ann Crum says,

> Few would argue that practical experience in theater production is essential to a dramaturg's training. Without it we run the risk of breeding what James Magruder calls "the killjoy dramaturg," who, ignorant of the process of making theater, may know what to say, but says it two weeks too early or three weeks too late. Timing is everything, as are subtlety, irony and wit." (Crum 70)

Being aware of the delicacies of timing and of rehearsal room protocols demands an investment in the material conditions of theatre attained through close observation of the production process. Though much can be learned during the process, the dramaturg who has spent time observing the ways that theatre is constructed is considerably better placed to understand how he or she might figure collaboratively in that process.

A PUBLIC FACE FOR THE SHOW

Most dramaturgs are required, as part of their work (particularly for professional productions), to give lectures and/or participate in talkback sessions with the audience. How formal such events are, and how much structuring of his thoughts the dramaturg therefore needs to do in advance, varies tremendously, as does the level of sophistication of the audience for such talks. To function effectively in these situations, the dramaturg needs the kind of expertise demanded by the roles detailed above, coupled with a thorough knowledge of how the show evolved in rehearsal (and, perhaps, in performance) and a facility for public speaking. This public face is also likely to be required in print as well, since dramaturgs tend also to contribute synopses, short essays, or other notes for the production's program or website. All these "performances" give the dramaturg the opportunity to further the production and company's goals through direct interaction with the audience, and can therefore be a surprisingly useful and connective engagement in the intellectual life of the community.

The dramaturg as I have presented her, then, is an invaluable collaborative resource, a source of expert opinion and information that most directors would be delighted to have. Certainly some of what a dramaturg does crosses into the realm of the director, but the same might be said of a costume or lighting designer. Actors, far from being mere directorial mouthpieces or puppets, shape a production by what they bring to the rehearsal and performance process, however much a director may arrange and channel their talents, training, and experience. Indeed, since theatre is a fundamentally collaborative endeavor and is not authored by a single individual (such as the director), this is true of everyone involved in the production, regardless of how minor their contribution or how invisible they are to the audience. Can a good director do the job of a dramaturg? Possibly, as a good director could probably be his own stage manager or sound designer, but the director's function is to shape the entire production by utilizing the talents and creative energies of all those involved. To have to focus on a particular aspect of the show (an aspect that could be better handled by someone more suitably prepared who can give it all her attention) takes the director away from his larger focus. The impulse for directors to absorb the role of the dramaturg (though sometimes presented in perfectly reasonable terms such as the desire to familiarize oneself with the script through the editorial process) can problematically position the director as auteur, potentially eroding the medium's essentially communal mode of artistic construction.

While few people make a living as dramaturgs, dramaturgy is a full-time job requiring complete immersal in the texts and contexts of the plays themselves. Directors rarely have the skills, opportunity, or expertise to function as scholars, historians, and critics, particularly when they have their directorial duties to handle at the same time. This is why they have dramaturgs. This is why they *need* dramaturgs. The work contributed by a good dramaturg might never be visible to the audience, but it invariably makes for a stronger, more intellectually coherent, show.

THE DRAMATURG'S STATUS

Different companies and productions use dramaturgs in different ways and give them varying degrees of visibility. Some will prefer the dramaturg to talk only to the director, others assume a value to giving the dramaturg access to actors and members of the production team. While the more involvement the dramaturg has the better (and involvement is usually linked to access), the dramaturg is often placed in a difficult position with regard to the larger chain of command. This is compounded by factors such as a director's lack of familiarity with this particular dramaturg or with the role of dramaturgs in general, or by major disagreements on the nature of the show, and such problems can be significantly worsened by the dramaturg's own actions and attitude.

In dealing with problems of communication within the company, or in wrestling with frank disagreements between members of the show's staff, the dramaturg should remember that he or she ultimately owes allegiance to the show and the company in its broadest sense. This allegiance takes precedence over other allegiances to the text, to the director, to one or more of the actors, or to the body of beliefs and ideas that the dramaturg may bring to the process. All the dramaturg's work, all discussion, all disagreement should ultimately be in the service of the production. The dramaturg that subverts a show because he does not agree with certain choices within it, or advises the actors to contradict the director's wishes, is serving his own agenda and not that of the show. On the other hand, the dramaturg who passively assents to ideas and choices he believes to be misguided for the sake of supporting a director or an actor is also doing a disservice to the production itself. The downside of collaboration is that while everyone involved can take credit for good work, they are all—dramaturg included—responsible for bad work. Letting things go without a fight can be as damaging to the show as active subversion.

The dramaturg's position is complicated by the fact that theatre practitioners of all sorts are sometimes uncertain how to use a dramaturg and how much weight to give her opinion. Since dramaturgs are often considered adjunct positions, actors are often unused to thinking of the dramaturg as central to the process of constructing the show. Some dramaturgs encourage such an opinion by confining themselves to those aspects of the show that are the most academic (program notes, audience lectures, and the like) and which tend to feel the most peripheral to the production. As dramaturgs play a larger and more engaged role in the staging of Shakespeare, this perception will inevitably change, though it takes time. For now, a dramaturg has to win over the trust of those he or she is working with, something that may not happen until the show is due to open and may not happen more fully until the dramaturg has worked with the company several times. The only way to encourage the sense of the dramaturg's centrality to the process is to act as if it is already true, by involving oneself in every suitable aspect of the show from the earliest opportunity, and by being constantly on hand throughout the rehearsal process to serve as a resource for the production. The dramaturg who shows up for the first or second time during tech and starts giving notes will be met with understandable wariness, even hostility.

It is not merely the uncertainty about where the dramaturg stands in the company's pecking order that can create a sense of opposition within the show's staff. The dramaturg is a uniquely intellectual presence on the production, and it is easy for him to become shrouded by a cloud of academic dust. The awareness of this cloud promotes a sense of the dramaturg as removed from, even irrelevant to, the material and emotive matter of the show, matter that is (in our generally anti-intellectual culture) often considered to be the real meat of the production. While I think such a perspective has to be countered and shown to be simplistic, the dramaturg does not effectively erode such perceptions merely by vociferous arguing from within his cloud, or from behind what Bruce Evers—one of GSF's longest serving actors—calls his "musty tomes." The dramaturg must, like all theatre people, have a sense of the performative and cannot simply rely on knowledge and intellect. Theatre is, after all, about communication, and the dramaturg must be prepared to modulate her communicative mode depending on the situation.

I learned an important lesson in one of my first dramaturgical assignments when the director (Sabin Epstein) told me in no uncertain terms to stop taking notes during rehearsal. I was surprised and frustrated, and I retreated temporarily from the process, feeling unappreciated. But he was right. The room was small, the actors only a few feet from where I sat beside the director, and my constant scribbling was distracting, making the actors uneasy and self-conscious. The director was not trying to shut me out of the process, only to remind me of its fragility, a fragility I was threatening by being overzealous. A dramaturg comes with a certain amount of "expert" insight that, rooted as it is assumed to be in text and history, can easily create the sense of an oppressive and restrictive presence which I alluded to earlier: the dramaturg as Shakespeare police. Theatre is a delicate structure that requires openness and spontaneity, things that clumsily insistent intellectualizing can seriously jeopardize if not sufficiently modulate. The dramaturg must, therefore, do all that is possible to convince those working on the show of his collaborative investment in that show rather than his policing of it.

Another consequence of the dramaturg's curious status is that even after a show has opened, his work—unlike that of most of those involved—is largely invisible to the audience. Even the cast who have worked with the dramaturg on a daily basis may think of his contribution as relatively minor. This, alas, goes with the territory. Theatre is very much about what one can see or hear in the actual performance, and much of that which has gone toward constructing that performance—if it is not clearly visible in the performative moment itself—tends to get elided, even forgotten. No audience can mistake an actor's contribution, and the director is still largely assumed (albeit problematically) to be the production's ultimate author. The builders and designers of sound, costume, lighting, props, and sets have the products of their labors out there in the public eye. The dramaturg does not. Even if the audience knows what a dramaturg does (and most don't), they are unlikely to really see or hear the work consciously, and when they do, they tend to ascribe it to one of the people (like the actors or director) whose work they think they understand.

This is not a problem in itself, except that it is all too easy for the dramaturg to fall into the trap of believing her contribution to be minor, even nonexistent. I have spent many days in rehearsal where I have felt that I have had no measurable impact on the production. I have often experienced the frustration of making a real contribution to a particular moment, only to see that moment dropped or watch it morph into something else before the show opened. The difficulty, of course, is that much of the dramaturg's work is in details, and it is all too easy to lose sight of details when presented with the finished product, which is so large and (hopefully) impressive, however much we remember—as we always should—that the sprawling monolith which is the show is composed and held together with details.

Actors don't always know how much conversation takes place between the dramaturg and the director outside the rehearsal room, or how much of a hand the dramaturg had in building the script, and once they are in the rehearsal room they have to be so focused on their own work that they can also fail to notice the extent to which directorial suggestions may originate in the dramaturg, or how a fellow actor's performance may shift after a brief chat with the dramaturg in the corner. Again, the dramaturg's work is often about details, and once an actor has started to grasp his or her part, the mechanics and origin of such details tend to get suppressed in the actor's mind as he or she attempts to "own" the role. I don't want to overstate this, of course, or imply that the dramaturg is secretly responsible for everything of consequence in the show, but it is certainly true that the dramaturg's somewhat marginal status (a person with one foot in the world of theatre, the other in the world of books, history, and scholarship, a person always somehow on the edge of the production) makes it all too easy to forget or to minimize her real contributions.

Consider, for example, the script. Few things have a more palpable effect on the show itself, but an audience or critical review will rarely comment on the cutting of the script because few people have the knowledge and expertise to do so effectively. Even the actors forget how opaque and baffling unedited Shakespeare can be once they have a handle on what they are doing in a scene. But people do hear the dramaturg's work, even if they don't know what they are hearing. When audiences commend a production for its clarity, for example, the ease with which they could follow the plot and character arcs of the story, they think they are complimenting the actors, director, or voice coach (and, of course, they are), but they are also complimenting the dramaturg whose work to make the script intelligible on paper and then in the mouths of the actors can make or break the show. That few people will realize how much of this was done by the dramaturg is both unfortunately predictable and of no real consequence. The dramaturg's allegiance is to the show, not to his own ego, so the sense of having done good work that has shaped the production however minutely, however indirectly, the sense of having helped to get—and keep—the thing on its rails, should be its own reward. Yes, it can be maddening to feel you have put weeks of your life into a production and then be largely unacknowledged for it, but key people (the director, actors, audience members who really know the play) usually do recognize your work, and the longer the dramaturg works with a company, the more he will be trusted and recognized for his contributions. In time, of course, I think that dramaturgs will be both more common and more valued as the nature of their skills become better known. For now, comfort yourself with the idea that when a production flops, no one blames the dramaturg!

Section II

Dramaturgical Issues:
The Theory

The purpose of this section is to raise some of the core debates that have a direct bearing on the practice of dramaturgy. For readers coming from a background in performance theory, much of what follows will be familiar, though I have tried to inflect each issue so that it sheds light on expressly dramaturgical concerns. While there are times when a dramaturg might have to address these issues directly—in conversation with a director, for example, or in response to an audience member's question during a talkback—much of this theoretical material will not be an overt or even a visible part of what the dramaturg says. I would go so far as to say that a lot of it does not belong in the rehearsal room at all, where heady and abstract discussion can often be off-putting to actors who have to make concrete choices with their voices and bodies. I prefer to keep such matters for discussion either in the formal setting of the lecture, or the very informal setting of sharing thoughts with the company over lunch or a drink.

What these issues do provide, however, is a foundation that, though not performed for everyone in the company, helps to shape a vision of the dramaturgical function and of how the production itself might be constructed. The principles shaped by these broad theoretical ideas and debates should help the dramaturg to become clearer in his own mind about where he stands on issues of great significance for the production and the company, helping him to give intellectual cohesion to the assumptions and methodologies of the show as they are manifested by its material choices. Editing the performance script, for example, which is perhaps the most important work the Shakespearean dramaturg can do, requires a thorough knowledge of how the text of the play, which can be bought in a store, came into being, what it contains, and upon what it is based. Editing also requires a clear theoretical sense of the relationship of that text to the performed event whose script the dramaturg is preparing. Theory clarifies methodology, gives it a logical underpinning against which the dramaturg can test his assumptions and impulses, and those of everyone involved in or watching the show. Without a grasp on such issues, the work can become random or inconsistent, it can fasten itself to spurious ideas, which may, in turn, produce serious flaws in the production. It can leave the dramaturg unsure of herself and unable to offer the kind of expert advice that directors expect and need. A good sense of theory breeds confidence, opens up bolder, more interesting possibilities, and helps to root not just the dramaturg but the production itself in a clear, interesting, and intellectually defensible position.

1

Theatrical Collaboration and the Construction of Meaning

It is a commonplace to assert that all theatre is collaborative, but in our consideration of how the dramaturg fits into a production, it bears repeating. Theatre is what happens when the combined efforts and talents of one group of people (the company) interact with another group of people (the audience). This first group varies in their degree of visibility to the latter, usually with the actors being most visible (and some actors will be, of course, more visible than others), though after the actors the order of the list is arguable. For some, the director is paradoxically most "visible" because he or she is seen as somehow "authoring" the production. While directors do have a major hand in shaping the production, often—but not always—giving it its distinctive feel or agenda, they do not truly author the show in any meaningful way. This is not simply because other people's work contributes to the effect of the show, though this is certainly true (costume and set designers are also fairly "visible" to the audience, sound and lighting perhaps a little less so, depending on the show). The production can have no author in the usual sense of that word (a single creative origin) because the show constructs meaning in the air of performance. Meaning, and by that I include everything the audience is aware of, even if it is not intellectual, is built from moment to moment within the show and is as effected by what an actor looks like (the color and style of his clothes, his brand of handsomeness, race, stance, or overall "presence") as much as by what he consciously does or says at the director's bidding. Yes, a director selects, but much of what the show finally contains is not the result of deliberate or even conscious choice and much of what is consciously chosen originates in people other than the directors. The director gives her approval to the work the people in the company bring to the show, but this does not make the director the show's author.

The theatrical experience is, moreover, semiotically reflexive: not a broadcast but an exchange.[1] In other words, meaning is constructed by the audience as they watch, but their presence and their mood also plays to the actors on stage, thereby altering the theatrical event. This is why no two performances can ever be precisely alike, because however tightly the onstage action is structured, it will always be affected by the behavior of the audience.

Theatre is not film (which does not alter according to the mood of the viewer), and it is this essential fragility that makes it electrifying and polymorphous. An audience that laughs loud and early or at the wrong times, an individual sleeping in the front row, a cell phone ringing, all alter the rhythms of the show and change the nature of the performance just as a dropped line, missed light cue, faulty prop, mistimed entrance, or new and spontaneous piece of stage business alter the show. In a crucial sense then, the audience is not simply a passive receiver of meaning, but has a hand in its construction.

The collaborative nature of theatre is thus not simply about the inability of any one member of the company to claim authorship of the show. It affirms that theatre is essentially the activity of a community, and this is as true in the performance as it is in the rehearsal hall. With this said, it becomes a good deal easier to underscore the extent to which the dramaturg should embrace the collaborative conditions of the genre and, in turn, be allowed to function as something other than a mere academic extra, a walk-on part allowing the show to tip its hat to scholarship while ensuring that the dramaturg remains peripheral to the actual production. Rather, the dramaturg should be welcomed as bringing a specialist knowledge and insight comparable to that of the other contributors on the production staff.

Collaboration, of course, demands not only that the dramaturg be given a real voice in the production, but also that she should not expect to enforce her opinions on the production. The first step to reconciling oneself to such an idea is to fully recognize that any play is open to myriad interpretive methods, and that the production need not replicate a single reading in order to be accurate or authentic. For dramaturgs trained in literary criticism this can prove a difficult prejudice to overcome, since despite the theoretical (especially deconstructive) emphasis on the indeterminacy of an absolute meaning rooted solely in the text, literary academics are still accustomed to building arguments based on the words of the text, shaping its disparate and often ambiguous or contradictory content into a univocal position. Coming to a production with this "thesis-driven" approach to the play needs to be modulated, therefore, not simply because the text can be read in many different ways and with broadly different interpretive emphases, but also because the text is not the sole or originating locus of the production's meaning, an idea to which I will return in chapter 2.

For now let me simply assert that the collaborative nature and spirit of theatre does not merely *permit* the dramaturg's involvement in and influence on the production. Rather the dramaturg's very marginality, her slightly indeterminate status as "between" categories, means that she does more than partake of that collaborative spirit: she must embody it and be its constant advocate. This is more than a matter of deferring to a director when disagreements arise; it is about fostering a spirit of communicative openness and exchange, something that extends beyond the rehearsals, even beyond the show proper, to a sense of inclusivity and mutual education in dealing with the audience. This is especially important in theatres that are anchored by a particular community, where a spirit of collaborative mutuality goes far beyond the making of a good show, and affirms the place of the company in and of the community, deconstructing all those familiar elitisms about theatre in general and Shakespeare in particular.

2

THE TEXT/PERFORMANCE RELATIONSHIP

While it is self-evident that non-textual factors (costume, set, light, and so on) contribute to the meaning of a production, it is still widely assumed in some quarters that the text is the sole starting point of the production and that the non-textual elements are there to somehow facilitate the transfer of meaning from the text to the stage. This is not an adequate explanation of the relationship between text and performance. In the following two subsections I shall briefly suggest some of the reasons why.

THE INCONSTANT WORD: RENAISSANCE PLAYS IN PRINT

Since a good amount of the dramaturg's job hinges on matters of text, particularly on the evaluation and interpretation of extant versions of the play, leading to the construction and subsequent editing of a performance script, it is important that we consider the nature of the Renaissance play as it was originally printed. This is especially pressing since there is a vogue (often, interestingly, in theatre companies that do not employ dramaturgs) to use early texts, particularly the 1623 Folio, as the bedrock of performance practice.

The plays of Shakespeare and his contemporaries frequently exist in more than one early form. A standard modern edition of a Shakespeare play will combine elements from all known early versions printed in or close to Shakespeare's own lifetime, selecting those parts the editors think to be the best or most authorial in instances where there are competing versions of the same lines. This text is subsequently modified by a tradition of Shakespeare scholarship and editing practices, built on debates and choices made in the intervening centuries, most of which (in a good edition) will be signaled in textual notes. Finally, the present editor will make new choices. The result is a conflation combining early but disparate textual elements with informed guesswork, all of which—however discretely—shape the nature, aesthetics, and content of the play. The result is, unavoidably, at a certain remove from the text as originally printed and performed, and one of the few things that can be said with certainty about the modern edition is that it is not the same as that which was staged in Shakespeare's own lifetime. Contrary to some assumptions, however, this fact does not undermine the use of such texts as the basis of "authentic" productions on the stage today.

What evidence we have from the original period suggests that the play texts were fluid, subject to alteration, correction, and addition. These modifications were sometimes made by the authors, but they were probably also made by the actors themselves, and we know of several instances in which other authors were paid to revise or write "additions" for a play that was being revived.[1] Plays were staged in a rotating monthly repertory, and when they became stale, they were replaced by other plays. A certain demand for newness (something endemic throughout the period with its burgeoning capitalistic sensibilities) drove the theatrical marketplace, and it thus seems likely that the plays would evolve during their initial run, as they almost certainly did if they were subsequently revived.

We have none of Shakespeare's original manuscript drafts of his plays—usually called foul papers—written in his own hand (except perhaps a disputed fragment from a collaborative play called *Sir Thomas More*). We have neither the polished version of these papers (called fair copy), which included emendations such as improved stage directions, nor the book of the play, which was based on the fair copy and which the theatre company used as the prompt book and the text that had to be approved by the state censor (the Master of the Revels). In other words, we have no copies of the plays that represent either original authorial draft or the versions that served as the core of the first theatrical productions. The earliest texts we have of the plays are in published editions, the status of which varies tremendously from play to play.

Shakespeare's plays were first printed either in single volumes called quartos (or, in the case of *Othello*, in the smaller octavo format) or collected in the first "complete" edition of Shakespeare's works (the First Folio), which appeared in 1623, seven years after his death, having been compiled by some of his friends and colleagues. This complete edition reproduced many plays attributed to Shakespeare for the first time, and without this volume and its subsequent reprintings, eighteen plays that had not already been published would almost certainly have been lost. The printing of plays in quarto seems to have been a fairly haphazard practice, and the texts themselves (unlike the much more expensive folio) were largely considered disposable literature. There is considerable scholarly debate over why some plays were printed in early quartos (usually close to their date of composition or first staging) and some weren't, and even more debate about the extent to which the author was actively involved in their printing.

In the case of some early texts—sometimes referred to as "bad quartos"—like the first printing of *Hamlet* (1603), the gap between the play in this form and in more polished subsequent printings (in the case of *Hamlet* in 1604–5, 1611, and one other dateless quarto that appeared before the 1623 Folio) has led some scholars to conclude that they were "memorial reconstructions," printings based on as much of the play as some of the actors could remember and sold to make a swift (and illegal) profit. Other scholars argue that these so-called "bad quartos" represent acting versions of the play for use on provincial tours. Both these arguments have recently come under fire by scholars claiming that the "bad quartos" reflect a version of the play that was performed in London; the other, more polished versions (like the Second Quarto of *Hamlet*, for example) reflecting a version of the play that was written specifically to be printed.[2]

Some scholars argue that Shakespeare had no particular interest in the printing of his plays, and that the early quartos which seem to contain "good" texts were printed only because the company needed to raise money quickly, perhaps because they could not perform in London at the time since the playhouses had been closed for fear of plague (a recurrent problem throughout Shakespeare's career).[3] This is why, these scholars argue, the Folio was not compiled until after the author's death, why it was put together by his friends as a posthumous (and potentially profitable) monument, rather than something in which Shakespeare himself had any direct hand. Ben Jonson had published a similar collection of his dramatic "works" in 1616, containing lengthy and annotated versions of his plays aimed at a sophisticated reader, and had been taunted for his hubris. Plays, say these scholars, were not the stuff of serious literature. Shakespeare wrote for the stage and what happened to his plays afterward, he did not care.

Other scholars, however, think that Shakespeare and Jonson had more in common than has been traditionally assumed, and that Shakespeare had expressly literary aspirations.[4] This is why there is such a difference between the "bad" and "good" quartos, the former being an acting version, the latter being a carefully constructed and revised version aimed specifically at a print market. The Folio, such critics argue, thus reflects plays that have been at least in part reshaped by the author for publication, even if the final editing and collection was not completed until after his death. One of the problems with the Folio text, and with some of the early Quartos, is that many of the plays are simply too long to have been conveniently performed in the approximately two hours that seems to have been standard for a Renaissance performance. While it is probable that the dialogue moved faster in those pre-Stanislavskian days, it seems at least as likely that the plays were subsequently enlarged and amended for publication. This seems especially true of the tragedies that tend to be longer than the comedies and which, it is argued, lent themselves more naturally to a sense of plays as "serious" printed literature.[5]

Whatever one believes about Shakespeare's attitude to print, both positions problematize the idea that the plays as they exist in these early printings represent accurately what was originally staged. If Shakespeare did revise for print, then the early printings are probably significantly different from the version that had been acted, and if he didn't, then the final polishing and shaping of the plays was left up to friends, printers, and others. Indeed, if Shakespeare didn't write for print, we are left with the sense that what was only a *version* of the play has been rendered constant and static in ways it was never intended to be, the shifting and frequently altered script arbitrarily accruing the permanence of the book.

One instance of this artificial permanence lies in the extent to which the plays were shaped by the actors themselves, not just by the business of their bodies, but by their words. We know that there was a strong improvisational tradition in early Renaissance comedy, for example, and though that fell out of favor during Shakespeare's heyday, it remains uncertain how much of the plays as they were performed incorporated clowning that was not scripted. Hamlet's much discussed advice to the players would not include the remark that the clowns should say no more than was set down for them, if the practice was not a fairly common one.[6] It is even possible that improvisation of this kind found its way into the printed version of the plays, whether it was approved by the author or not. Conversely, we know from a few printers' notes that the printers themselves sometimes made changes, removing, as was the case in Richard Jones's printing of Marlowe's *Tamburlaine*, for example, comic scenes that were considered tonally inappropriate.

We must further acknowledge that many of the plays show signs of other authorial hands, some welcome (collaborators, the insertions of actors, and the like) some less welcome (censorship, for example). Scholars like Gary Taylor have made compelling arguments for the hand of Thomas Middleton in *Macbeth* and *Measure for Measure*, while Shakespeare's contemporary Thomas Heywood positions his play *The English Traveller* "amongst two hundred and twenty, in which I have had either an entire hand, or at the least a maine finger" (A3). This collaborative method seems to have been a basic condition of playwriting in the Renaissance, but present day critics and audiences tend to think that something penned by more than one person— particularly if the additions or changes were made without the original author's consent—is less "pure." In fact, of course, such valorizing of the author's individual genius, particularly where drama is concerned, does not fully develop conceptually until the nineteenth century, so any quest for authorial purity in matters of Renaissance theatrical texts is likely to prove elusive.

Whichever side of these issues the dramaturg finds herself, certain principles become clear. There is and can be no master text for a play by Shakespeare. All the printed versions that exist from the period in which Shakespeare was writing seem provisional, subject to change. In some cases, these changes are minor, but in others they are extensive. The First Quarto of *Hamlet* is half the length of the Folio copy. The folio and quarto versions of *King Lear* are so thoroughly different (and the Quarto is not considered a "bad" quarto) that the Oxford Shakespeare series has opted to print them as separate plays; the story is the same, but the actual line to line expression is too extensively different (and in ways that do not clearly privilege one version over the other) to justify a simple conflation. Again, the one thing that can be said with certainty about all conflated texts (which is the vast majority of Shakespeare editions on book shelves) is that they were never seen on the stage or read in that form in the Renaissance.

This last idea has led to some theatre practitioners going back to the earliest printed versions for clues to a *truer* text and, moreover, clues as to how to play it. Some directors and actors swear by the First Folio, not just for the words it prints but for its punctuation and line spacing, which, they believe, contain hints for breathing, character psychology and delivery. There is no harm in such a belief, except where it screens out other possibilities (the First Folio is also dotted with manifest errors), particularly as there is no evidence to suggest that the Folio was ever used as or based upon acting editions. Rather, it seems that the Folio represents the most "writerly" form of the plays, and was thus aimed primarily at readers, not players. It might also be argued that matters of punctuation, capital letters, and so forth—elements that actors and directors are tempted to imbue with significance—are notoriously erratic in the Renaissance and would have been largely at the discretion of the printer (different printers seem to have particular leanings in matters of spelling and punctuation), so relying on such details as indicators of how to act a line seems doubly problematic. The argument that shifting speech prefixes (when a character's title shifts, say from "wife" to "lady" or "mother" as in *Romeo and Juliet*, or "Puck" to "Robin" in *A Midsummer Night's Dream*) seem to indicate the haphazard quality of the printing, not—as is sometimes argued—clues to how the character is to be played from scene to scene.

Acting approaches based on the quartos are less common (at least in part because half the Shakespeare canon does not exist in quarto), though there has been a rise in interest in performing the "bad quartos," which have, perhaps, a better claim to being "actorly" than does the Folio. Of course, such a practice is fraught with its own difficulties, not least of which is audience expectation. Most people who come to see *Hamlet* expect to hear "To be or not to be, that is the question," not Q1's "To be or not to be, ay there's the point."

This is not to say, of course, that consideration of the earliest printed versions of Renaissance plays is of no value. On the contrary, such texts provide one of the most useful tools to which the dramaturg has access, but they must be handled cautiously, without the assumption that they provide an unmediated expression of either theatrical event or authorial intent. Both are lost to us, a fact that is ultimately extremely liberating; it allows the dramaturg and the production in general to pursue an expressly theatrical experience while being informed, but not enslaved, by the letter of the surviving text.

From Page to Stage?

For modern productions, the text exists prior to the theatrical event because some printed version of the play can be read and used to design the show long before it ever opens. This does not mean, however, that the text somehow *contains* the show or that the show merely makes choices prescribed or offered by the text. Text and performance are different genres, and though performance can be read as if it were text (by reviewers and performance critics, for example), that does not render the two the same. In real terms the link between a script of *Othello* and a stage production with the same title are only slightly closer than the relationship between that same script and a production of *Who's Afraid of Virginia Woolf*? The reason for this is because the respective genres, books on the one hand, performance on the other, are formally different and convey meaning in radically different (even oppositional) ways. Yes, the words of the former become the dialogue of the latter, but this is not a process of simple translation in which the same content is conveyed by a method that is only superficially different.

A book or other printed text consists of words on a page. The reader interprets the words and constructs their meaning in her head based on her understanding of those words as shaped by her personal and cultural context. A stage production, by contrast, tells the story through a more kinetic experience of sight and sound. Unlike reading, for example, in which the reader can pause or reread a passage when she feels like it, the pace of a performed story is dictated by powers largely outside the audience's control. The appearance of the characters is not open to interpretation as they are in books because they are personated by actors. Audience response is shaped by mood created by lighting and music. The very meaning of the words themselves is radically altered by the actors' take on their lines. In short, it is impossible to move simply, as the old phrase has it, from *page to stage*. Too much of the final performance is non-textual in origin, and even that which is textual (dialogue, for example) is inflected by the prior and expressly theatrical work of the production. This is an inevitable consequence of the theatre's semiotics. On the page, the lines are open to an almost infinite set of possible deliveries. On stage (except in the most postmodern of productions where lines are done in more than one way!) a single choice—conscious or otherwise—is made.

One of the most common complaints critics make about productions and films of Shakespeare plays is the way they close off possibilities. Richard III, they say, is more than a Fascist dictator. Hamlet is not merely an Oedipal puppet. Such arguments are usually (but not always) made by people who disagree with the given interpretive approach, and their objections may be perfectly valid. What is not valid is bewailing a production's closing off of different readings through their performance choices. The infinite variety of *readings* (and I use that word advisedly) is something reserved for armchair criticism, not for theatrical production that of necessity must make a series of single choices.[7]

Consider, for example, Hamlet's opening line when he is addressed by his uncle/stepfather, King Claudius: "A little more than kin and less than kind" (1.2.65). This is typical of Hamlet's riddling utterances and contains several puns, particularly in the movement from kin to kind, with that latter term's double suggestion of "type" (he's suggesting that he is a different "kind" of man than his uncle) and "kindly" (he's suggesting the ill will he feels toward his uncle). The king's response ("how is it that the clouds still hang on you?") is not clearly a response to Hamlet's remark, and many productions will thus have Hamlet deliver his line in a way that Claudius does not hear. Hamlet's second line, however, which is just as riddling and potentially provocative ("Not so, my Lord, I am too much in the sun"), is more clearly dialogue. Someone reading this exchange on the page is thus permitted to imagine a range of possibilities, not just about the words Hamlet emphasizes and the meaning he thus suggests, but about whether or not the first line is dialogue or aside. If it is dialogue, how playful, respectful, or defiant might the line be? If an aside, then addressed to whom? Himself? The audience? Someone else on stage? Moreover, the moment is simultaneously affected by other non-textual factors that take us steadily further from the lines. How do the other characters on stage react? Are they even paying attention? Is Hamlet sitting or standing? With what attitude of body? What is he wearing? Is his appearance formal, polished, or slovenly? Is he in the center of the stage or at its margins? On the stage all of these are possible, but unlike the armchair reader for whom all the possibilities can exist concurrently in his head, the performance can only deliver the line in one way, before moving on to the next and the next, each utterance being made out of the myriad possibilities, building on all which has gone before until, at the end of the show, we have a single and unique version of the play culled from the range of possibilities suggested by the text.

To extend the sense of difference between page and stage, we must also acknowledge that the factors which determine the choices the production makes are also not necessarily textual. This may seem like a small matter, but a Hamlet who mutters to himself is very different from one who chats to the audience, and different again from one who makes fun of or otherwise defies the king. A minute choice about the delivery of a single line can thus have radical implications for the rest of the play, and in a play like *Hamlet* the depiction of the central character will have extensive impact on every aspect of the production. The problem is that while all the above possibilities can work with that line (and with the character who develops in the remainder of the play), there is nothing in the text itself that demands the part be played any one way, as there is nothing that says whether the character should be short or tall, blond or brunette. Critics still like to argue whether Hamlet is genuinely mad or sane, and the range of possibilities is extensive for any given moment in the play (if one assumes that his madness or sanity fluctuates as the play progresses), but nowhere does the text say definitively which is the "correct" reading. Not only must a production make a series of choices when presented with every single crux in the play (however minor), the factors that inform each of those choices are as likely to come from outside the text as inside it. The actor playing Hamlet will have certain preferences, but they will have as much to do with who the actor is as a person as what is in the text. The truth of this is made clear when one reads accounts of productions from fifty years ago or more, where the text is assumed to say directly things that we find almost impossible to consider today. As readers and performers, we are the products of our cultural and historical moment and the sum of our personal experiences. When we perform a play, we bring to it assumptions and points of focus that may connect to the text but originate outside it, which is one of the reasons that even if we did have an authorial manuscript that was clearly better than any other subsequent text, we wouldn't be able to read it as the original actors did.

That the material conditions of theatre and its practitioners significantly inflect any approach to the text has always been the case. On the Renaissance stage, the actors interpreted and constructed their performance, and without the help of a director in the way we understand that role. Sometimes the authors may have had a hand in "directing" the play, but this doesn't make Shakespeare the author of the production in the way he was the author of the play text, theatre, as I have said, being essentially collaborative. It thus has to be acknowledged that despite the claims from actors and critics that a production should somehow "do" the text (and usually this means "do the text *in the way I read it*"), this is simply not possible. To put the play on stage is to transform it, to draw on fundamentally different means of signification (the theatrical "language" of bodies and sound and light as opposed to the literary signifiers of print on paper), and render it new. Theatre is not a conduit for text, a telephone through which the textual essence emerges at the other end basically the same as it was when it went in, it is a wormhole, a rift in the space–time continuum through which one is transported to a radically different, strange, and—hopefully—wonderful universe. The text is not the production "in potential," it does not predict or even direct the performance, and it cannot somehow contain or restrict the number of "correct" stagings of the play, since too much of the production is determined by its own genre of communication (the theatre). In short, the production does not move from page to stage. A play (text) and a production are fundamentally different things, and while they are interconnected, the former does not dictate or originate the latter.

The implications of these premises for the dramaturg are, of course, far-reaching. The dramaturg has to understand the play as a textual entity, she has to be steeped in every nuance of its detail, she has to wrestle with that which is ancient or foreign in its language in order to make sense of it, but she has also to recognize that the text is crucially provisional because it can neither determine the production nor be seen as the sole fountainhead of the artistic product that is to develop on stage. A central premise thus emerges: all production is adaptation.

3

ADAPTATION AND AUTHENTICITY

The idea that all production is adaptation makes the purists cry foul and can create a sense of panic in even the most liberal of theatre practitioners. But if, as I have argued above, the shift from page to stage is a shift in genre, a shift into a different mode of signification in which non-textual elements come into play, come even to dominate the production's textual elements (e.g., the words of the dialogue), the idea is unavoidable. The only question for debate is one of degree, not kind. How adaptive can a production be, for example, before it ceases to be Shakespeare at all?

Different theorists will, of course, answer this in different ways. For some, the question is irrelevant because the notion of not being Shakespeare is an implicitly textual idea, one that has been rendered moot the moment we accept that production is neither a translation nor a consequence of text. To say that a production has deviated from being Shakespeare means that it has deviated from a textual entity we are used to calling "Shakespeare" but which was itself fluid and has no claim to authority over the theatrical event, as Shakespeare the man of the theatre would have understood.

My answer to the question then has more to do with audience than it does with pure theory. For example, Tom Stoppard's *Rosencrantz and Guildenstern Are Dead* is heavily informed by *Hamlet*, but is its own play with its own title, despite borrowing characters, plot, and lines from Shakespeare's original. It couldn't exist without that original. But what if Stoppard had produced the play (working less as a playwright and more as a dramaturg or director), not as a new work inspired by an old, but a new take on a familiar play? What if he had staged it not as *Rosencrantz and Guildenstern Are Dead* but as *Hamlet*?

Adaptation, though unavoidable, is a slippery slope. If we cut *Hamlet* (as we surely must), by a few hundred lines, it is still *Hamlet*. What if we cut out Ophelia? What if we write extra scenes? What if we were to make two minor characters the stars of the show (say, Rosencrantz and Guildenstern)? I do not believe there is a measurable point that the play ceases to be *Hamlet*, but I think that staging *Rosencrantz and Guildenstern Are Dead* and calling it *Hamlet* is disingenuous and misleading. Different readers and audience members will disagree on the most important elements of the *Hamlet* story (how crucial, Rosencrantz and Guildenstern are, for example, depends on what you think the play is about), but few will dispute the centrality of the title character. Where Shakespeare is concerned, as I have already said, language is also crucial to the value of the play, so a version of the story that abandons Shakespeare's language altogether (like, say, Disney's *The Lion King*), while still telling a version of the story, is not doing Shakespeare in any way an audience will generally accept. This is very important because the dramaturg, though invested in the text and invested in theory, is crucially invested in how things will play to an audience. I believe that Shakespeare has always to be new on stage, but newness at the expense of what the audience came to see must be handled with great caution.

In my treatment of script preparation in this book, I will be taking what I consider to be a middle ground, neither "purist" in its sense of the script nor radically adaptive. People who prefer their theatre to inhabit one end or other of that spectrum will doubtless see my suggestions as either disruptive of what makes Shakespeare great or dully conservative. I accept this because I feel that both of these contrary positions are in some ways untenable. The "purist" approach to the text seems to me theoretically and historically muddled, while the radical, heavily adaptive approach strays close to the production of new work (like Stoppard's). Since all new work is implicitly in dialogue with the past, let us announce a new play's newness and be done with it.

The idea that Shakespeare's worth is solely tied to his words is, of course, problematic, since Shakespeare is a cultural force, albeit an amorphous one, interwoven with the politics of art, education, and culture in its broadest sense, and some productions may want to engage these ideas by radically altering the language of the play. That said, it remains my contention that the core of the plays is their words, and to completely rewrite them, while still being potentially "Shakespearean" in the broadest theoretical sense (one that recognizes "Shakespeare" as little more than an authorizing fiction), will be seen by audiences as, at best, counterintuitive, at worst, false advertising.

This of course fudges the question of those productions that do not throw out the Shakespeare play altogether, but still take a fairly heavily adaptive approach. One way around this is to herald a degree of change in the title. We know *Macbett* is by Verdi, not by Shakespeare, though there are clear relationships between the two. We know that Baz Luhrmann's recent *Romeo + Juliet* is aware of its adaptive status and that this is being signaled to the audience before they take their seats. Again, I think the audience should be your guide, always accepting that the production is supposed to be creating a moment of theatre first, and adhering to a notion of Shakespeare second. While there is no measurable point at which a heavily adaptive show ceases to be Shakespeare, a production should always be true to the range of adaptive vision it sets for itself. In all but the most postmodern of productions, this requires a unity of perspective on the relationship between text and performance, a constant (preferably one established earlier in or before rehearsals) that sets up the poles between which the show will move. While I will return to a consideration of how theatre authorizes itself in chapter 7, let me say here that the question of when a production ceases to be "authentic Shakespeare" has more to do with the goals established by the show itself than it does with adherence to a monolithic notion of Shakespeare, since all theatrical production is necessarily the construction of a new work of art, not merely the revisiting of an old one.

Recognizing the inherently adaptive nature of theatre, furthermore, connects productions to the spirit of the age that produced the plays first. There is growing evidence, as I have said earlier, about the way that Renaissance theatre inhabited a curious, liminal, and innovative space, on the borders of conventional cultural entertainment as the theatres were on the borders of the city of London itself, either pushed out among the taverns and whorehouses of the South bank or preserved from the authority of the city fathers in the old church lands within the cities (the "liberties"). The theatres functioned between high literary culture and the lowest forms of popular entertainment such as bear baiting. They embodied a spirit of community, which was inventing the rules, decorum, and logic of itself as it evolved. The theatres were themselves adaptive, changing to meet the tastes and demands of their new audience, and something of their spirit is essential to dramaturgy today, particularly when Shakespeare on stage can seem so dull, so archival in spirit. Some dramaturgs believe it is their job to protect the text from the production, but the converse is at least as valid, perhaps more so. The spirit of immediacy and adventure that was so central to the Renaissance stage is all too easily lost, and one of the surest ways to kill it outright is to be overly restrictive in dictating what a production can or cannot do with a text. I say do "with" a text and not "to" it, because the text always survives the production. It remains for sale in the theatre lobby, and if discovering precisely what it says is of prime importance to the audience, then they can always be directed there, to read.

"As It Was Originally Done": The Logic Behind Historical Reconstruction

Given this problematic relationship between text and performance, it is not surprising that some companies seek to authorize their productions in other ways. One of the most common is a reversion to Renaissance theatrical practice or, more commonly, something that superficially resembles it. Such an approach often works from the assumption that to modernize the manner in which the play is staged, or to otherwise pursue an overarching "concept," is to impose a frame on the play that dilutes it or renders it somehow impure. What I have already said should have made clear some of the reasons I find such assumptions unsound, but since the impulse to stage supposedly Renaissance productions of Shakespeare is such a common one—and something audiences often seem to want—the issue is worth more careful scrutiny.

Let us assume for the moment that we can get access to a historically authentic text of a Shakespeare play and that we want to re-create the way that play was originally staged, on the assumption that getting close to the performance conditions under which the play was first performed will make its meanings resonate more fully, more accurately. What elements of the Renaissance stage must we rebuild in order to recreate these performance conditions?

The simple and brutally destructive answer to that question is, of course, *all of them*. If a production seeks to claim a purity of approach based on historical method, then it does not make sense (and seriously undermines the legitimating authority of the production) if the show itself turns out to be palpably inaccurate in its recreation of history. This is a significant problem, since the recreation of Renaissance theatrical conditions is close to physically impossible.

It is not enough, for example, to put the actors in doublet and hose and call that a good approximation of the Elizabethan or Jacobean stage. A Renaissance production had numerous features that are at odds with present day theatre. The physical structure of the theatre itself, for example, is something few modern companies can recreate. The amphitheatres were open-air spaces with thrust stages, each with a character of their own; since the dimensions and nature of the house and stage varied in important ways from one theatre to the next, the experience of playgoing at the Globe was significantly different from that of the Rose or the Swan. They held 2,000–3,000 people, perhaps as many as 800 standing, a fact that significantly affected the theatrical dynamic since the day-lit space meant that all these people were clearly visible to the actors on stage.[1] Lines given to the house meant making eye contact with the audience. Without a darkened auditorium, then, the actor–audience dynamic was fundamentally different from that which evolved later, and which today is still largely premised on the assumption of an invisible fourth wall that clearly did not exist on the Renaissance stage.[2]

The stage itself was practically bare, though the actors used its physical nature (dimensions, columns, doors, discovery space, raised gallery, and so on) in their performance. To create a historically accurate production, the recreation must do the same. All the actors on stage were male, the heroines played by boys, older women probably played by men. There were no special lighting effects, and sound effects were generally limited to what could be accomplished with fireworks, cannons, and the primitive equipment used to simulate the noise of weather, ocean, and so forth. Music was played on instruments such as lutes and hautboys (a species of oboe). The language was uttered in a dialect we would find foreign today, even baffling.

So far so good. These are things that, though expensive and requiring considerable research and labor, a company might be able to re-create if they wanted to. But things get more complicated thereafter. What, for example, about acting style? The rehearsal period of a Renaissance play was probably measured in days, not weeks or months, and the actors generally received only their own parts. Some present day companies have occasionally re-created this practice and have found it instructive (the actors who often remarking that it forces them to listen to the other people on stage much more closely than when they are familiar with every line in the scene). But how does a modern actor approximate what his (not her) Renaissance contemporary did when they actually started to act?

We know that the repertory system generally meant that a company actor had to know his part to about thirty plays at any given time and that new plays were introduced monthly. This raises real questions about the degree of depth that an actor could explore in his part. We know also that there were stock gestures (some derived from the physical mannerisms taught to accompany oratorical performance) and even symbolic postures or movements that announced something relevant to the audience as when the player queen in *Hamlet* is described in the stage direction as making "passionate action" or when Ophelia enters "distracted." We have no accurate record of exactly how actors performed in this period, but everything we know about theatre thereafter suggests that the acting was more declamatory and less naturalistic than what subsequently evolved. Even if (and it is a considerable *if*) the contemporary actor could recreate what his Elizabethan or Jacobean counterpart did on stage, how would a contemporary audience react? Would they see an acting style that was suggestively different and rooted historically, or would they see acting that was merely bad?[3]

This is a vital and illuminating question because it reminds us of where we began this discussion. Meaning is not determined by the text, nor is it determined by the intention of the actor or director. Meaning is constructed in the moment of performance by the audience, a fact that creates a significant— even an insurmountable—hurdle for those wishing to do Renaissance Shakespeare. Because however meticulously we re-create the original theatrical conditions of the Renaissance stage (using, for example, precisely the same kind of oak for the framing structure of the building—as the new Globe did in London—and the same archeologically re-created plaster to fill in the walls), however much we dress our all-male actors in meticulously researched Renaissance costume, we CANNOT re-create a Renaissance audience, and this is where the project (insofar as that project is solely about re-creating the way the plays first created meaning on stage) must inevitably fail.

The Renaissance audience was a unique collection of people, one whose socially diverse nature has rarely been seen since in a theatre. They drank, sat on the stage, talked among themselves, and heckled the actors. They were people, moreover, who were invested in their own historical moment because they were of that moment. How they reacted, for example, to the presence of the cross-dressed boy actor, to what extent they saw male or female in that actor's performance, is a matter of heated scholarly debate.[4] What can be said with certainty, however, is that they saw it differently than would a modern audience unused to such a convention, and this is surely true of every aspect of the Renaissance stage. In Renaissance dress they saw the specifics of their period, which were nuanced and contemporary. Presented with the same costumes, a modern audience sees something quite different, something vaguely old, slightly odd, but rendered safe by its distanced antiquity. In the language of the plays, they heard not the difficult and fusty terminology of the theatre and the schoolroom as we tend to, but the language of their own lives. They heard the puns and quips and flashes of political satire, the topical references to religious debate and geographical discovery, the use (or abuse) of rhetorical methods learned in school, the parodies of contemporary poets we have long forgotten, the professional languages of law and medicine, the terminology of war and horticulture, fashion and treason, and these things resonated with them because they were the stuff of which their universe was constructed. When a modern audience hears such things we do not hear it as they did. Much of it we miss, much of it resonates but in ways that are different from the ways it resonated originally, and even what we comprehend through study and immersal in the period we comprehend only through a glass darkly, as we might learn of a culture at the other side of the world, the details of which we can grasp intellectually but will never be able to comprehend on a gut level because we have not been raised in that culture. Even if we were able to go back in time and watch these plays as they were originally staged, we would not see them as their original audience saw them because we can never be that original audience.

The Renaissance audience did not think about the same things we do, did not even think of themselves in the way we do. Most historians of the period would contend that the audience did not merely perceive the theatre differently from us, they perceived the world differently. Their very bodies were different from ours because they thought about them differently, saturated as they were with ideological pressures of church, state, economics, the vestiges of feudalism, and so on. Such things shaped their sense of self in ways that rendered their brand of subjectivity significantly different from ours. Consider, for example, how Renaissance standards of hygiene, the absence of flushing toilets, the infrequency of bathing, the fact that many people would own only two or three complete changes of clothing, the lack of anything resembling modern sanitation or plumbing would alter your sense of self, and by that I do not simply mean how close you sit next to people in the theatre. The way we perceive our bodies shapes our priorities and the way we view the world. To take a more extreme (but related) example, how is one's sense of self altered by a world in which death is so immediate, in the mutilated body of the criminal displayed to public view in pieces, in the astronomical infant mortality figures, or in the frequent and devastating ravages of bubonic plague? How is your sense of self altered by the assumption that you are unlikely to live more than forty or fifty years? How different was early modern subjectivity, particularly since it was defined by the structures of power that enforced a degree of control even in matters of thought with which we are today unfamiliar.[5]

Beyond the material difficulties of recreating Renaissance theatrical conditions then, there is a deeper theoretical problem that raises a question about *why* we should pursue the doomed attempt to re-create the original staging conditions. The kind of modern subjectivity we assume today was in its infancy in the Renaissance, and what we like to think of as universal, as binding us through a common humanity, is often the product of our current cultural mindset. A society that forbad the appearance of women on stage for moral reasons cannot, for example, conceive of gender or of romantic love in the ways we do. When we stage Renaissance plays "as they were originally done" then, we enact what is at best a kind of fiction, and one that generally performs a species of dubious cultural nostalgia. It does not re-create the past in a way that allows us to see the plays as the original audiences did because that is not possible; what it does instead is create a sense of something that is familiar—in that it accords to our sense of the iconic Shakespeare anchored in the past—and something unfamiliar—in that it announces its distance from ourselves, thus at least potentially disarming the plays as energized by and connected to the present. Both are, to my mind, so deeply problematic that they leave me wondering why anyone would go to all the trouble to produce something that is finally likely to feel so fusty and irrelevant.

That said, there is nothing innately wrong with Renaissance (or, more often, pseudo-Renaissance) settings for the plays and they can have certain advantages in their perceived harmony between dress and word: a soldier in such a production does not run into the problem of saying "rapier" while

brandishing a machine gun. The problem comes from the idea that such gestures toward a (largely spurious) historicity somehow make the production
better, purer, truer. They don't. Audiences may make that assumption (as
they may assume that Shakespeare's lines should be spoken with a British
accent, despite the fact that modern American speech is as close to
Renaissance English as modern British, probably closer), but that does not
mean it should be pandered to. A Renaissance setting for a production might
be a good and productive choice, but let us not forget that it is a choice, and
as such no better than any other.

5

The Nature and Use-Value
of History

The above skepticism about the attempt to reconstruct Renaissance theatrical conditions in no way renders obsolete the use of history by the dramaturg. On the contrary, historical research is essential to understanding how the plays worked—or may have worked—in their first days on the stage and this is, in turn, vital for the dramaturg's evaluating and reshaping the plays for performance today. Some basic principles must, however, be borne in mind throughout.

History is not a sequence of facts that produce a harmonious and totalized narrative. Contrary to the way many of us were taught, history is actually a web of interwoven discourses that are shaped by competing ideological positions, many of which contradict each other. Truth as a hard, knowable, and precise entity is elusive at best, and probably doesn't exist in any form that we can agree on. Instead, truth is vague, imprecise, contested. Indeed, from a certain perspective there is no historical Truth, only truths.

Consider how difficult it is to make definitive statements about relatively recent history, about the intentions behind certain political or military acts; we are used to the idea that there is disagreement, sometimes violent disagreement, about exactly what happened and why in, for example, the Vietnam or Gulf Wars. Part of the reason for this disagreement is that discussion of matters as charged with feeling and political import as war means that most reporting of the subject tends to be inflected, however subtly, with the ideological bias of the person doing the reporting. The more we investigate a historical event, the less clear its details become as we are forced to engage the opinions of witnesses who disagree about what they saw or heard and what it all meant. The further back in time we go, of course, the more dubious becomes the idea that the actual facts of the past can be reconstructed precisely.

Nevertheless, it is only relatively recently that history has engaged such matters directly, and in some circles the idea that the past is knowable, measurable, and precise is still alive and kicking. In the study of Renaissance drama, for example, there has long been an understanding that we comprehend the plays better if we comprehend the period in which they were produced, but early-twentieth-century accounts of the place of Shakespeare in history were marked by a flawed sense of what history itself was. One of the common consequences of this is emblematized by a question such as "What did the Elizabethans believe about . . . ?" The question does not need to be concluded, because nothing that follows those first few words can be answerable in a way that will satisfy the spirit of the question.

The Elizabethans. The assumption behind the words is that all the people alive in England while Elizabeth was on the throne (from 1564 to 1603) were somehow the same, or at least thought in the same way. It may be that there was a greater consensus of belief on some issues in this period than there is today, but the virtual impossibility of making any accurate statement about what twenty-first-century Americans (all of them) believe should give us pause. There is no question that Shakespeare's audience were different from us and that their opinions differed from ours, but to assume that they all thought and felt the same way is bad logic and worse history. It is tempting to reduce a culture that existed 400 years ago to simple statements about what used to be called The Elizabethan World Picture (after Tillyard), but it isn't accurate and, again, closes off fruitful avenues of inquiry for a dramaturg and the production on which she is working.

Take, for example, the matter of social stratification in Elizabethan England. There is no doubt that Elizabethan society was significantly structured along lines of rank and of what came later to be called class, and that these structures affected not just the opportunities of the people, but their behaviors and thought processes as well. One of the dominant metaphors for this heavily stratified worldview was called the Great Chain of Being, a view of the cosmos, theology, the animal and vegetable kingdoms, and the social world of humankind that placed all things in a ranked hierarchy supposedly dictated by nature and the Creator. The fusion of theological and social hierarchy led to the widespread assumption that marks of rank were God-given, and fed the Medieval notion of the Divine Right of Kings, a philosophical position made much of by Elizabeth's successor, James I. Evidence for the extent to which people accepted this worldview include documents such as the Homily on Obedience, which was read in Anglican churches periodically and which asserted the essential nature of rank and of behaving—and thinking— in accord with one's status. The Elizabethans, the argument concludes, loved and accepted an order based on rank, thus they venerated their monarch and preserved a sense of obedience at every point on the chain of command in ways that emulated what they believed to be the relationship among God, angels, and beasts.

The problem with such a view, of course, is that the testimony provided by a document such as the Homily on Obedience is only one view, an official view, and one that is contradicted by the frequent instances in which people violated that sense of rank, for example, by marriage to someone outside their class, or by disputing the testimony of one's social superior, or by armed rebellion against the monarch. With such events and their surrounding discourse taken into account, the Homily on Obedience begins to look not like proof of something that was beyond challenge, but like an official attempt to enforce something that was pointedly *not* taken as read.

Rank was supposedly self-evident, marked by an individual's carriage, spending power, clothing, and so forth, but the shift toward what we might call proto or nascent capitalism was changing all that, giving some of the people who had been close to the bottom of the ladder only a generation or two before the opportunity to buy and sell their way into the ranks of the gentry, even the aristocracy. At the same time, some of the older aristocracy were finding themselves so financially strapped that they could no longer maintain their position in society.[1] The resultant sense of social flux was a source of some anxiety in the period, and one of the ways that the government attempted to regulate such social mobility was through laws that regulated what people were permitted to wear in public. Only men of certain rank could carry a sword, certain fabrics were reserved for people who could demonstrate an income over a certain amount, and so forth. Again, rather than demonstrating the fixity of social position in the period, sumptuary legislation seems to demonstrate the opposite: that the world was in flux, and rank—now more clearly linked to spending power than bloodline—was up for grabs. Moreover, the fact that these laws were passed frequently throughout Elizabeth's reign further suggests that they were virtually unenforceable and were having little or no effect on what people were actually wearing.

But if history has to acknowledge a gap between official discourse and actual practice in matters of dress, imagine how much more difficult it is to come up with statements on what people *thought*. The Elizabethan settlement of religion had made England Anglican, but to what extent did such a statement go to the hearts of a people who had, until very recently, considered themselves Catholic? Certainly Catholics who wore their faith on their sleeves were severely punished, but—as the recent example of the Soviet Union demonstrates—religion that is oppressed does not simply disappear. Some scholars believe, for example, that Shakespeare's own family, and possibly the playwright himself, retained their Catholicism privately, while others argue that many people were completely oblivious to the finer points of theology and could not report even the basic principles of their Christian faith.[2] Again, a sense of ideological commonality—let alone homogeneity—seems woefully out of step with the facts.

My point here is not, of course, that nothing can be said about Renaissance history accurately, rather that history is not single and univocal, but polyvocal, a set of competing discursive positions and communities vying with each other for mastery. It thus becomes nonsense to contend that Shakespeare's plays represent "what people believed then." Even if the plays can be reduced to a single set of attitudes on issues such as monarchy, the relationship between the sexes, religion, social mobility, and so forth, those ideas are in dialogue with the community of which their author was a member. They don't represent the beliefs of that community any more simply or directly than a popular movie reflects exactly the ideas of its entire audience.

By the same token, we cannot say that the plays are constantly in opposition to Renaissance culture and its ideologies, as if they were written by someone who somehow existed outside his society and culture. This is not possible. Shakespeare was a man of his time, and his plays therefore reflect the culture around them, though they do so in complex and shifting ways, playing the role neither of simple adversary nor of mirror. Artists like Shakespeare do not float free of their historical context by virtue of their genius (as is sometimes argued) creating work that is timeless and universal in its applications. Rather, the raw material of their work is specific to their context, and we must therefore do what we can to recover that context in order to understand how the work was produced and in what particular ways it might have originally resonated with its audience. There is nothing that is clearly universal and timeless, since everything (love, hunger, desire, God) is ultimately culturally mediated, shaped by the specific moments in which people live; we must therefore acknowledge that Shakespeare is not our contemporary, his audiences aren't ours, and while Elizabethan lives and beliefs can sometimes look and sound very like ours they are, finally, not the same, cannot be the same by sheer virtue of what makes them Elizabethan. What we consider to be timeless or universal is generally a projection of feelings or ideas of our own onto a world that, removed in both time and space, would find them quite alien. In this respect, history is like geography: it divides. The ghost of old Hamlet plays differently in Africa because the culture framing the audience's reception is different. With universals and transhistoricals abandoned, the only way to understand a work of literary art, then, is by saturating oneself in the culture that first produced it, however diverse and multifaceted—even contradictory— that culture seems when examined historically.

Having said that, a central tenet of the New Historicist approach is that the very factors which insist that we understand artistic work historically (particularly the extent to which our culture makes us different from that work's first audience) also suggests that we can never fully grasp the historical moment in which the work was produced. Many scholars suggest that our very subjectivity renders us different from our Renaissance forbears, which means that we can never wholly comprehend their exact historical situation because the way we perceive and reflect upon the world is itself a product of our own cultural moment.[3] History—in the sense of the actual recovery and comprehension of the past—is thus a theoretical impossibility, and all we can hope to do is rediscover ways of reading the past that have been occluded or ignored.

For the dramaturg, this raises a difficulty. On the one hand, history is the only way to understand the plays, but on the other, the past is lost to us, and it and its people are alien to us. Furthermore, since theatre is about communication in the present rather than through the (theoretically impossible) reconstruction of the past, we must acknowledge a species of disconnect between the ends of historical (usually literary) scholarship and the production of live theatre. History can inform our sense of how the plays may have once been received, but such information cannot be used to determine present day production directly.

The best use of history is consequently indirect, treating history as a particularly fruitful form of analogy. Some aspect of the play resonated in a particular way because of specific historical conditions; these conditions are now lost to us, but our own cultural moment may supply an analogous idea or point of contact that we can use to ground the production. If we understand the implications of a certain phrase in the context of its original utterance, for example, we might be able to connect that phrase to something in the present, something to which it does not directly refer by virtue of a transhistorical universality, but to which it might be legitimately linked by the authorizing power of the theatrical moment. The result is not, of course, the same as was manifested on the Renaissance stage, but there is a relationship between the two that preserves something of the play's original life while redrawing it for the contemporary audience. History is thus usable, but it must be mediated, used by the dramaturg in collaboration with the demands of the present.

6

Ambiguity and Polyvocality
in the Plays

I have argued that history is not a progression of facts but a competing series of discourses, and a similar idea might be applied to the content of the plays themselves. People are understandably eager to find definitive messages in the plays, statements of authorial opinion, or cumulative expressions of support for one particular social, political, or philosophical ideology. The plays have, however, a slippery quality that makes their appropriation by a single ideological position both extremely easy and, if we want that position to be uncomplicated and univocal, extremely difficult. Since Shakespeare became a monument of high culture he has been used by the proponents of wildly divergent ideas as if his works clearly and unproblematically manifest these ideas. Shakespeare has been a Marxist, a Fascist, a royalist, a republican (in every sense of that word), a Democrat, a Feminist, an anti-feminist, a racist, an exposer of the moral and cultural flaws of racism, and so on. The very plurality of these readings—and the extent to which they have been held up as incontrovertible—suggests their speciousness. Indeed, it is the plays' ambiguity, their openness to interpretation and to refashioning on stage, that is partly responsible for their continued presence today.

The plays are not novels or treatises. They consist almost exclusively of dialogue—which is never simply thought (even soliloquy which—particularly on the day-lit Renaissance stage, suggests a performative dimension)—without a clear authorial voice telling us what we are supposed to make of the characters' words. This is the problem with that impulse to reduce Shakespeare to pithy or elegant expressions of perspicacity, something that renders all his characters mere authorial mouthpieces. "To thine own self be true," says Shakespeare. "Kill all the lawyers," says Shakespeare. But it is not Shakespeare who says these things, it is his characters, and his characters (as is palpably the case in those two examples) are almost always flawed, shaped by their predicament, mislead, self-deceiving, or merely stupid. The fairly obvious sense that we should not expect unmediated wisdom from fictional characters who are enmeshed in a story much larger than themselves, should also make us wary of expecting any given play (or the Shakespeare canon as a whole) to produce single strands of thought on any given subject, strands that point clearly toward a definitive opinion. The dramaturg must thus be wary of holding onto any single reading so firmly that any alternate reading starts to feel simply wrong.

This sense of pluralism and polyvocality is not some wishy-washy, postmodern principle that says we merely have to respect anything and everything that everyone says. It is hard-wired into the nature of the dramatic text, which is as full of space as it is of words, and into the theatrical process that adds to these words the light, color, and living bodies that help to *write* what the production says. The plays are inevitably fluid and open to broadly divergent approaches on stage because the text alone does not dictate where meaning in theatre is constructed.

Such an assumption, however, runs contrary to much literary criticism, which attempts to identify a single, cohesive reading of a play and argue it with reference to the text so that any alternative reading is seemingly rendered invalid. The papers written by students and the articles written by professors frequently demand this thesis-driven unity in the plays, and it is all too easy to assume that such unity, rather than being imposed upon the text by the essayist for rhetorical purposes, is actually innate to the play. Theatre is not the writing of academic papers through other people's bodies, and the dramaturg who goes into the rehearsal room with such an attitude is likely to be surprised and disappointed.

As Paul Yachnin has argued, the political impact of *King Lear* in its own day was probably negligible, and how its political content would then have been read is uncertain at best.[1] What is clear is that Shakespeare has become a locus for the production of political meanings, a locus whose power is tied to his status in the world of arts and ideas (and, in Yachnin's view, one tied to the marketing of deluxe goods to a popular audience). Shakespeare *can* be extremely political on stage but this does not suggest that the text of the play has a clear political message, rather that the complexity and flexibility of the play lends itself to being given a particular political (or philosophical, or theological, and so on) shape. The dramaturg should thus go into initial discussions about the production with an array of possibilities to hand, and while it is perfectly appropriate for her to give greater emphasis to some than to others, insistence that the production reproduces a particular ideological position on the grounds that the text can only be read in those terms is to overly limit the production and, paradoxically, the text itself, which is rarely clear-cut on anything. To push for a largely univocal production based on one's own ideological bent is, of course, a different matter entirely.

Polyvocality is not, of course, reserved for the play text itself, but is inherent to the mechanics of theatrical production. The nature of theatrical semiotics says that even though a production makes a series of individual choices from one stage moment to the next, choices that could potentially unify or clarify ambiguities or dissonant voices and ideological structures within the script, this in no way renders the production itself univocal. Meaning, as I have already said, is created in the air of performance, constructed (not merely received) by the audience, and we must acknowledge that the audience is always plural, being composed (hopefully) of multiple persons. These persons will all read the show differently, based not on "objective" assessment of the action (which cannot truly exist), but on political predilections, mood, preoccupation, sightlines, personal associations with a character, attraction to an actor, and the like. Indeed, even a single audience member has to be considered a plural presence since his or her mindset or degree of attention is bound to fluctuate slightly from moment to moment. The result is a semiotic entropy produced by the theatrical equivalent of the butterfly effect, where tiny and seemingly inconsequential stage action can massively alter the overall audience response in ways that are diffuse and unpredictable.

The dramaturg's sense of the play must then take into account (insofar as is possible) the fact that "the audience" will read things differently according to their own experience and mood, and this must be factored into how the show will attempt to create its meanings. Much of this sense of audience as erratic or uneven does not produce much that is useable, of course, but through the deliberate adoption of different viewpoints, the dramaturg can approximate some audience response.[2] For example, broad demographic issues should be factored into any consideration of how certain moments will play, particularly where they intersect with especially charged issues. A largely white cast staging *The Tempest* with a black Caliban, for example, must be prepared for how black audience members might read the show's attitude to race. Similar issues saturate Shakespeare's work, both of race (consider the potentially incendiary issues in *Othello*, *The Merchant of Venice*, or *Titus Andronicus*, for example) and of gender (all the plays contain potentially problematic representations of women). Class issues, though less inflammatory, likewise may spark unintended alliance or hostility from the audience if not adequately anticipated beforehand. In short, even the most balanced and deliberate production will spark audience argument about what they actually saw and what it meant, but if the production staff embraces the inherent polyvocality of the production (as well as the text) and thus enters the process forewarned, a higher degree of semiotic control can be retained.

Such awareness should also teach the dramaturg humility, since the most rigorously thought-through approach to a play must inevitably depend on the vagaries not merely of its execution on stage, but of audience response. If nothing else, a sense of the fragility of the theatrical dynamic and the inherent inability to force a univocal reading by an audience should make the dramaturg more open to collaboration and less insistent on a single approach to the play.

7

AUTHORSHIP, AUTHORITY, AND AUTHORIZATION

I want to use this chapter as way of pulling together the different ideas and approaches I have used thus far. The conclusions I come to here, conclusions underwritten by W.B. Worthen's book *Shakespeare and the Authority of Performance*, provide the authorizing logic behind what I advocate in the more practically inflected part on the dramaturg's work before, during, and after the rehearsal process.

Critics such as Jonas Barish have identified what they call an "antitheatrical prejudice," a recurring set of objections to a kind of moral ambiguity that people in broadly different times and cultures have perceived to inhere in the very fabric of the theatrical experience. Much of this antitheatricalism has been bound to religious organizations or individuals, as was the case in the English Renaissance, but such religious concerns seem to be merely a part of a larger anxiety about the theatre's associations with falseness, dissimulation, and lying. Such charges have been leveled against all forms of fictive art, of course, but theatre seems to draw more than its fair share of such criticism, apparently because it transcends the storytelling of poetry or novels through the bodies of the performers. In short, the *enacted* status of theatrical storytelling has often been seen as claiming a kind of reality or truth that belies its fictional nature. In the Renaissance, for example, antitheatricalists verbally assaulted the players at the Rose and the Globe, not simply for the moral abuses they were perceived to encourage in their stories and in the fact that they drew people away from work and church, but for their quasi-divine acts of (false) creation, in which base-born players were made into gentlemen, kings, bishops, or women, by virtue of the theatrical medium. The audiences, they said, were encouraged to believe things (matters of plot and person) that were not true, to hear speeches made as if they were extempore by actors who had in fact memorized them earlier, actors who could manipulate the audience through appeals that were inherently false to their mind and heart. Theatre, they said, was a lie.

Such a position seems extreme today, but versions of it abound in a general anxiety about the authenticity of the theatrical experience. Audiences are constantly asking whether the show they are going to see will be "real"

Shakespeare. They make a point of reading the play before seeing the production, as if to alert themselves to any textual deviation, they worry about lines that may not be by the Bard himself, always seemingly preparing for some upcoming exam that will test the accuracy of their sense of the play as derived from the show against some Platonic ideal, some leather-bound, tome-like, and *definitive* text. What seems to have happened—at least where Shakespeare is concerned—is that those general antitheatrical anxieties about truthfulness have taken on a new specificity that is tied to the author's status as a poet and philosopher whose iconic position in the culture as a whole means that people are invested in knowing Shakespeare at least in part because such knowledge speaks of, even helps to shape, their own value and status. In such an environment where familiarity with Shakespeare accrues— or is perceived to accrue—a kind of cultural capital, all those old antitheatrical prejudices loom especially large. How can an audience be sure that the Shakespeare they are getting on stage is the real thing, and if they can't, how does that erode their investment in the theatrical event?

To reduce audiences at Shakespeare productions around the world to peo- ple seeking to advance their sense of inner worth through an appropriation of high culture is, of course, neither fair nor accurate. I do think, however, that an element of such a quest lurks in many who go to see Shakespeare on stage, and it is one of the reasons that that sense of the author as poet and philosopher (both, incidentally, suggestively untheatrical notions of the playwright) con- tinues to be so hard for the stage to escape.[1] In one of the most important books to touch on this subject, W.B. Worthen deconstructs the various ways that theatrical companies, practitioners, and productions have countered such latent anxieties and skepticism through various strategies designed to somehow "authorize" their work. The most common ways of doing this are through recourse to three pointedly untheatrical elements, the author, the text, and the historical past. As Worthen demonstrates, audiences, critics, and theatre practitioners of all stripes have been (and continue to be) at consider- able pains to underwrite the work of the theatre (and the financial metaphor is not inappropriate) with claims to authenticity that seek to sidestep or otherwise devalue that which makes the theatrical medium so unique: its flex- ibility, its collaborative nature, its productive (rather than merely reproduc- tive) essence and the fact that this essence inheres not simply in script but in light, music, and bodies, and so forth. In short, theatre has a tendency— disingenuously or otherwise—to mystify the core of what it does, what it is, in order to reassure audiences that what they are getting is somehow *real* based on criteria found outside itself.

In service of such an agenda, actors and directors often talk about discovering—or trying to discover—the author's intent as they work, as if studying the script or otherwise exploring the play will reveal Shakespeare's own coded presence, announcing once and for all what the great man thought and believed, as if replicating such thoughts and beliefs on stage make the show somehow more true (less like theatre's habitual lies), more *authentic*. This pursuit of authorial intent is not limited to the theatre, of course, but is especially odd in the theatre, a medium that is ultimately about the *creation* of art, not merely the reading of it, and creation in ways that require the creative energies of large numbers of people, not merely the author of the text. This pursuit of the author—like that more literal and even more ludicrous pursuit that seeks to "discover" the "true" author of the plays (Francis Bacon, Christopher Marlowe, the Seventeenth Earl of Oxford, and so on)— is both futile and misguided.

No examination of the text, in all its forms, will reveal Shakespeare's mind as no analysis of any art object can fully reveal the mind of the author, even those more clearly "authorized" genres such as novels, and no performance (be it historically "accurate" or contemporary experimental) can claim to access the playwright's intentions because those intentions died with him. What remains is text, which is fundamentally different from the person who created it—or partly created it, or created it in collaboration with someone else, or through the less direct but no less insistent collaboration of his own period and culture.

We can jettison the author, then, as an authorizing element in theatre, if only because there will never be anything like consensus about what the author intended, as the wildly different statements on what Shakespeare did intend over the centuries have made clear. The text itself is less easy to jettison, but it is perfectly clear that theatre is about more than words on a page; if it were not, every production working from the same text would be exactly the same. We must also jettison the sense that the authentic production is a historical one, one rooted in what the play once meant, since theatre is not a museum art form, but one uniquely wedded to the communicative moment in the fleeting present. The question of a production's "fidelity" to text, author, or other extrinsic "legitimating" concerns is rendered moot by the variety of productions which all claim that fidelity, a fact that, as Worthen says, "dramatizes the extent to which the assertion of authority is a fully rhetorical act, absorbed in the register of ideology" (18).

In short, theatre, like jazz, authorizes itself. It is not wholly dependent on the text, that text's author, or the period in which that text was produced, though it is clearly related to all three. It is dependent on its own internal logics, its own integrity, and on the singular collaborative semiotic exchange that defines it. Moreover, theatrical production is not measurable merely in terms of fidelity (to authorial intent, to text, to history) but is intrinsically linked to the production of what Worthen (after McGann, Shillingsburg, Barthes, and others) calls "the work." The *work* is that authorial element which precedes *text* and is materially absent from the *document* that manifests the text, and can only be approached through interpretation of the text (which is all that is read, not merely the material document in which the text might be seen to reside). The work is the object of consumption, while the text is the field in which such consumption is attempted. The theatrical performance does not replicate the text through reading and interpreting the document in which the text inheres (the script), nor does it simply enact a *version* somehow contained within the text or within the work (as Philip McGuire suggests [Worthen 14]). Worthen contends instead that the stage produces the work itself "through the application of a historically specific, ideologically contoured instrument: theatre practice. The work that is produced onstage in any performance of *Hamlet* is part of the history of the work, but it is also inevitably a new work as well" (20). Theatre applies "a variety of historically discrete, conventional, and changing practices to the text in their production of the work. No production speaks the text in an unmediated, or faithfully mediated, or unfaithfully mediated way. All productions betray the text, all texts betray the work" (21). In this sense, the production itself becomes a text—something to be read—one that independently produces the (new) work as part of "an emerging series of texts/performances rather than a restatement or return to a single source" (23). Performance, Worthen says, always risks being seen as transgressive because the cultural tradition embodied by a work is forced to tell a new story, though this is only transgressive if one believes that there are alternatives which, he concludes, there aren't. The work and the tradition it represents can only be perceived in text/performance, and the past is thus constantly "being remade by—and remaking—the present" (191).

That theatre authorizes itself means that to rely on other loci of authority (the author, the text, and so on) limits the possibilities that are unique to theatre, hamstringing the performance in pursuit of a wholly illusory authenticity grounded outside the medium. The general rhetorical mode, however, in both critical and theatrical circles, has been, as Worthen demonstrates at length, to use "Shakespeare" the author as a way of authorizing or justifying, what is, should be, or should not be done on stage under his name. Actors, directors, and critics persist in claiming to discover authorial intent, which proves the validity of their work in ways flatly contradicted by the dominant theoretical sense of how theatre actually works; instead of "liberating an authentic Shakespeare" as they claim—something that cannot exist—they "consistently work to authorize their own efforts under the sign of 'Shakespeare' " (39). While some of this is a disingenuous use of the magically authorizing Bard to insulate a production against audience criticism using the terms and values that audience is most likely to find pacifying, a good deal of it is much less self-conscious. This latter condition, which is probably common, is troubling because of the way it so easily undermines or circumscribes the constructive energies of actual theatre. Worthen thus concludes:

> If we claim to derive all authorizing power from the author or his documentary traces, then, we reify Shakespearean drama—and the past, the tradition it represents—as sacred text, as silent hieroglyphics we can only scan, interpret, struggle to decode. We impoverish, in other words, the work of our own performances, and the work of the plays in making our world. (191)

But Worthen's conclusions affect not just the way thinking people talk about theatrical production. They have insistent implications for the actual staging of Shakespeare, implications that demand the attention of the dramaturg. For while Worthen's book has become a kind of given in academic circles dealing with extant Shakespearean performance, it is far from clear what the book's consequences are for the construction of a production that might be called Shakespearean. Worthen is less interested, he says, in what "works" on stage, than in "what we think is happening, or how we account for what is happening, when it seems to 'work' " (42).

The dramaturg, by contrast, has to be concerned with what "works," though the investment in the realm of practical theatre, rather than in its surrounding discursive theory, in no way permits the dropping of those inverted commas. What "works" on stage will always be a matter of perspective drawn on individually subjective, historical, cultural, and ideological assumptions, and broad agreement does not mean that what is perceived to "work" is in any way more authoritative than anything that is generally perceived not to. Much of what was heralded as theatrical genius in the eighteenth century would be virtually unwatchable by a modern audience, and I seriously doubt that that modern audience would perceive much of what Richard Burbage and company did on the Globe stage to "work." The dramaturg's sense of what "works" must always therefore be provisional, tentative, aware of its idiosyncratically and momentarily inflected nature, aware that one line reading seems better than another not because one is more "Shakespearean," but because of the myriad personal and cultural factors convening on and shaping the dramaturg's subjectivity.

As well as breeding a healthy skepticism about the rhetoric of authority in the dramaturg, this sense of "Shakespeare"—not the fount of authenticity but as something forever deferred, a wholly malleable concept to be evoked in justification—makes two points clear. First, that the dramaturg's allegiance is to the show, not to the text or to the author (whatever that would mean), a principle that immediately establishes a certain constructive and adaptive impetus as being at the heart of all the dramaturg's work, including the editing of the script. Part of the dramaturg's mission then, contrary to much of what is said about it, is to push the production out from under the author's shadow, a shadow often manifested by vaguely historicized assumptions about attitudes within the plays or by the letter of the text itself. As the production's intellectual presence (not the production's smartest member, of course, but the one whose sphere of influence is that of the show's intellectual dimension), the function of the dramaturg is to embody a sense of the theory manifested by people such as Worthen in ways that will embolden the production and get it out from that shadow and into the light. If the dramaturg can give reasoned voice to such ideas in concretely theatrical terms, the production might start to discover its own legitimating authority, thereby producing not pale shadows of the play as textual document, but something that—while in one sense part of the work—is also a new work.

Second, an awareness of Worthen's marshalling of all that dubiously authenticating "Shakespeare" (those quotations from actors, directors, and critics talking about how they discover the author's intent in the play, and so forth) should leave the dramaturg with a clear sense of her minority position.

While such monolithic notions as the legitimating authority of authorial intent should be subverted wherever possible (in rehearsal, in preshow lectures, and the like) such subversion should thus be done with caution. This is one of the reasons that my advice on script editing is often premised on a certain underhandedness, in that I advocate an adaptive strategy that has real effects on the text but which goes largely unnoticed by the audience, even to the extent of using quasi-historical words and phrases to mask a cut or alteration. One cannot subvert an entire cultural bias in favor of authorship in general and Shakespeare in particular with a brief essay in a program. Such attitudes are too deeply ingrained. So while I do think those attitudes should be met head-on wherever possible, I also think that serving the show means making it "work" as well as possible without drawing attention to its adaptive strategies unduly.

Dramaturgy is often about picking winnable battles, and this goes for battles with audiences and newspaper critics as well as with actors and directors. Certain questionable assumptions about "Shakespeare" can be eroded (minimally) by theoretically savvy lectures and writing, but the show itself should not be left carrying that particular can. The theory should inform what the show does and how it was built, but it should not necessarily be forced into the footlights itself, lest all those questionable monoliths be invoked like specters sent—however problematically—to worry at the idea that the show *really* is "Shakespeare." More can be accomplished by flying under the radar.

8

DIFFERENT LANGUAGES

As the discussion in chapter 7 demonstrates, there sometimes seems to be a fundamentally different set of assumptions at the heart of the scholarly community than there is in the theatrical community, and these different assumptions are manifested in the very words with which these communities argue about what they do. As is the case with all languages, these two discursive communities make perfect sense within their own cultural boundaries, but when they are brought together they can produce discord, frustration, and antagonism. The dramaturg must therefore become a kind of translator, someone fluent—or close to fluent—in both languages, and thus able to move between the two respective cultures in ways producing the maximum number of productive options with the minimum amount of hostility. To map such a strategy, I must first demarcate the lines of difference between the two communities and this will necessitate a degree of over-simplification for which I apologize in advance.

Most theatre practitioners—naturalistically inclined actors in particular—think of plays as being about characters who are motivated from within, and they therefore view the action of the play as largely constructed out of those individual characters' journeys and relationships. Because they see the characters as rounded, thinking selves that are analogous to the actors themselves, the nature of these journeys and relationships is often thought of in universal terms, the plays "working" on stage because their core themes are still relevant today. Actors ask about their character's motivation (thereby assuming a degree of control) and what they want. They are encouraged to approach a scene with the question "why not . . .?" (a version of Stanislavski's "magic if") thus demonstrating that—despite all possible rhetoric to the contrary—they see the text simply as a starting point, the performance ultimately authorizing itself by projecting character onto actor and vice versa, then adjusting them until the shadows match.

Most literary scholars on the other hand (and it is as literary scholars that many Shakespearean dramaturgs get their training), though they often thought in comparable terms a half century or so ago, have moved steadily toward thinking about the plays in largely ideological terms, the characters representing not rounded people so much as positions engaged in a combative negotiation, exploring issues of philosophy, sociology, and politics, less as thinking selves than as mouth pieces for the discursive community that produced the original work.[1] Unlike the actors' assumption of a certain universality in thinking about character, the scholarly community tends to ground the issues around which characters move in expressly historicist terms, particularly involving strategies of power tied to the social and economic conditions manifested within the play. A good deal of literary scholarship still tends to focus on the triumvirate of cultural studies: gender, race, and class. Scholars do not assume the characters are in control (so questions about wants and motives tend to be secondary at best), rather that they are moved by ideological forces at work in the broader society. Literary scholars ask "why" (not "why not") when thinking about what a character does, because they see their function as identifying causes for the action of the play as manifested expressly by the text.

These are, as I have said, over-simplifications, but there is enough truth in them to suggest the breach that the dramaturg is attempting to span. In very real terms, theatre practitioners and literary scholars do not speak the same language, and the difference in diction and terminology they employ reflects substantial differences in what they perceive the plays to be and how they function. The entire first section of this book, for example, uses academic arguments about scholarly debates that some theatre practitioners will believe to have no relevance to the practical business of theatre.

The task of the dramaturg is, as I have said, to act as a translator and—better—a negotiator, by which I mean not simply someone mediating between opposing forces, but someone looking to make both forces listen to each other in ways that are enlightening. It is sometimes said that in battles between directors and actors, the director always loses, either because he is forced to give in, or because the actors are forced to do things to which they will not fully commit. The place of the dramaturg is analogous to that of the director but is infinitely more precarious because the dramaturg's authority is so fragile, not simply in terms of her place in the hierarchy, but because the dramaturg has to be *given* that authority by the rest of the staff, particularly the actors and director. The dramaturg, in other words, only has authority so long as the company believes she does, and her capacity to affect the production (which, as David Copelin says, is not so much about *power* as it is about *influence* [Copelin 21]) is thus predicated on her perceived usefulness. If the dramaturg speaks in ways that are convoluted or baffling to the company, harping on about theory or ideas that seem tangential at best, she will be tacitly shut out of the process. No one will listen to her, and she may as well go home.

All dramaturgical suggestions, be they pieces of research offered to a designer, concept refinements given to a director, or text notes given to an actor, must thus be shaped with special care for the values, logics, and assumptions of the person receiving the suggestion. In other words, the dramaturg must learn to be theatrical, in that he must be prepared to modulate his suggestions according to his audience. Even so, dramaturgs sometimes discover that certain ideas which are central to literary or theoretical analysis cannot easily be explored on stage without a significant redrawing of the production as a whole, and even then, perhaps, will simply not play on stage.

An excellent instance of this kind of impasse is described by Cary Mazer with regard to his staging of John Webster's *The Duchess of Malfi*, in which the production's dramaturg attacked his directorial assumption that the characters were looking for a sense of identity. The dramaturg explained that there was "no such thing as identity in the modern sense of psychological interiority or individual subjectivity, at the time the play was written" (Mazer, "Rebottling" 293). A solution to the competing ideological approaches of dramaturg and director was eventually achieved by combining modern approaches to certain aspects of the characters with premodern approaches to other aspects. In the case of Ferdinand, the Duchess' strangely fragmented twin brother, the modern approached centered on the character's fears. The premodern approach focused on his desires, which were limited by the character's inability to step into the more progressive brand of subjectivity manifested by his sister. Tellingly, however, Mazer says, "little of this was explicitly discussed with the actor in rehearsal" (306). Rather, Ferdinand's premodern crisis was translated into playable actions and rehearsal exercises, the actor being "encouraged to respond to stimuli, to act upon needs and desires, and to fear the things that terrified him—all playable actions—without inventing a biography, identifying a childhood trauma, or psychopathologizing Ferdinand's character." Ingeniously, then, a literary and theoretical insight into the play was allowed to manifest itself on stage without disrupting either the actor's process or the production's intelligibility to a modern audience.

Mazer's example, of course, assumes that there is something of value in maintaining some connection to the premodern situation that first generated the play, something not all directors would agree with. As a theoretically inclined director, Mazer is something of a rarity in the field (and one whose writings about staging Shakespeare are invaluable reading for prospective dramaturgs), and many dramaturgs raising this kind of idea may well find themselves having to decide at some point to abandon it altogether. While Mazer was able to attain a theoretically inflected study of the character by largely circumventing the different languages of scholarship and academia (in giving the actor things he could play he avoided explaining the theory and then leaving him to find a way to stage it), there are many comparable situations in which scholarly discourse runs contrary to theatrical practice.

Take, for example, Kate's highly problematic speech at the end of *The Taming of the Shrew* in which she (ostensibly) renounces her feisty resistance to male authority, saying that the wife's role is to be mild and obedient to her husband. There are, broadly speaking, three conventional readings of this speech, all of which are playable on stage. One is to take her at her word and suggest that she has been "broken" by Petruchio's "taming," an unfashionable approach that tends, these days, to turn all but the most broadly comic of productions into tragedy. The second is to take the speech sarcastically, as a preface to further defiance from Kate (something that can be pulled off convincingly if the sheer length of the speech is emphasized in the performance: Kate still has her "voice" and thus her will). This approach was used by the New Globe's all female production in 2003, and resulted in Kate dancing out of Petruchio's reach as she went on and on about how submissive wives should be. The third reading is more gently ironic and implies a shared joke between Kate and Petruchio at the expense of the other couples: Kate has learned (mainly in the fourth act's sun/moon scene on the road) how to play, and knows that more fun is to be had by mocking others slyly through doublespeak than through the simpler rhetorical battery she employed at the beginning of the play. The result is a kind of private harmony that sidesteps the political implications of the speech, or exposes them obliquely.

These three readings translate easily onto the modern stage because they view character as continuous and psychologically complex even as they negotiate a political issue (gender relations and hierarchy). Historicist critics, however, may argue that a more "authentically" Renaissance reading is none of these, but hinges instead on an essential discontinuity of characterization, in which the Kate of the previous act is simply effaced by the resurfacing of the dominant ideological discourse of the period. In such a reading, her character is effectively suspended as the topsy-turvy license of comedy is replaced by the official doctrine of male/female relationships as manifested in numerous treatises from the Elizabethan period. Kate turns into a spokeswoman—and the boy actor's nonbiological "femaleness" might be considered a telling undercutting of the speech itself—for the patriarchal party line. The original audience, accustomed to the restoration of an official version of order at the end of a play, would accept this because they could read the conventions of the genre and did not view the characters as psychologically continuous individuals.

It could be that this last reading is indeed the closest to what would have been perceived to have happened in a Renaissance theatre, but such a reading is extremely difficult to talk about in ways that are usable directly by an actress (not usually a boy actor) approaching that speech today. Indeed, were an actress to try and perform such a step into officialese, the result may well be completely baffling to a modern audience, which would probably see not the character *discontinuity* of the English Renaissance but a failed modern character *continuity*. If the dramaturg were to raise this reading—and I think she should, if only to add to the options on the table—it should be done very early in the process, since any attempt to play it radiates beyond that single character in that single moment. A nonrealist sense of character such as this reading implies would probably need to underscore the whole production, and could not simply be imported into that one crucial moment without seeming to jar the rest of the show. Some kind of narrative or rhetorical shift could be signaled, of course, in other ways (through a change in light or musical underscoring, for example) so that the show somehow flagged the change of tack. An actress left to pull this off by herself, however, is being required to change the way the audience understands character and theatre in general.

Any dramaturgical note to actors that suggests their characters are not real people is likely to produce this kind of difficulty, assuming the actors attempt to use the note, which is unlikely. To avoid such a problem, I would say that any statement about character that is likely to contradict the basic assumptions that the actors bring to their sense of their roles (and these will, of course, vary according to the company, even to the actor) should be brought to the director before the cast ever get into the rehearsal room. That way the director can consider drawing attention to the issue through design elements or a particular style of acting, for example. If on balance the idea does not seem compelling enough to explore throughout the show, it probably should not be brought up at all thereafter.

Since we are dealing with differences in the language of theatre and scholarship, however, we should remember that some ideas can be made to cross over from one side to the other if they are adequately translated. To present an actor with a disruption in the continuity of their subjectivity, as in the above example, will probably produce panic, suspicion, or hostility, but many issues that could produce the same response if presented in the same theoretical language can be made playable if carefully phrased for the right audience. As one would talk to music and lighting designers in terms of mood, energy, and focus, so the dramaturg should learn to present her conceptual issues in suitable terms. If talking to directors, for example, frame the issue as how something might play to the audience. If talking to a naturalistic actor, rethink the issue and present leading questions about character goals, desires, and needs. In the case of *Shrew*, for example, talking in broad and theoretical terms about patriarchy will not get you as far as reducing that patriarchy to the specifics of how the individual character is affected by its details as manifested not by an invisible social order or hegemony, but by the specifics of the play (father, suitors, clothing, behavior, and so on). If there seems to be no viable way of talking about an idea except in theoretical terms, the dramaturg must gauge how receptive the target audience would be or be prepared to drop it. This is not about intelligence: theatre people think in concrete and emotionally immediate terms because their medium is concrete and emotionally immediate. Some things are only interesting and workable on paper. That doesn't invalidate them, it just means that they apply only to thinking about the play, not to staging this particular production. It is part of the dramaturg's task to recognize when an idea is impractical or at odds with other aspects of the show.

The dramaturg should also be aware that intellectual cohesion and theoretically rigorous argument will not always carry the day, particularly when other people in the company are already wedded to ideas or principles that are grounded in other kinds of thought which have become central to the show. Again, the dramaturg needs to pick her battles carefully. Some are worth fighting even when she knows she will lose because a point needs to be made for the record, but many aren't. Frustrating though it undoubtedly can be to feel right but ignored, the dramaturg should remember that her first duty is to the show, and vociferous insistence on ideas that probably aren't going to get followed can be a source of dissent that divides or otherwise undermines the production. The spirit of collaboration that gives the dramaturg a voice in the process also demands that, from time to time, she must let certain ideas go, however attractive or intellectually compelling they may seem.

WHY STAGE SHAKESPEARE?

All the agonizing in the preceding chapters (particularly in chapter 7) over what is and is not Shakespearean, coupled with an imperative toward theatre that is clear, immediate, and relevant to the current audience, raises the obvious question: why bother? If we want clarity and immediacy instead of stultifying Bardolotry and archaism, why not just stage modern drama? In pretheoretical days, when Shakespeare's texts and their value in terms of content seemed more stable and self-evident, such a question would have seemed stupid or blasphemous, but a book such as this assumes that a production has much to do to make the plays "work" on stage, so some sort of answer—however personal and tentative—seems in order.

The trick, of course, is to be able to answer the question (why bother staging Shakespeare?) without recourse to the usual dubious smoke screens—such as universality and transhistoricity—while identifying those things about the plays that make them worth doing, even as we assume that the play texts do not contain or define their own substance on stage. Yet we must begin answering such a question with the words. It is the words on the page that are taught and quoted, and it is on them, rather than the inherent theatricality of the plays, that Shakespeare's cultural reputation tends to rest. The beauty and power of the plays, their density and insight—be it psychological, political, or whatever—tends to reside at the level of individual utterance, rather than in larger points of plot. However foreign that language can appear to some audiences, it remains the root of the plays' value; I am constantly struck by words and phrases that seem to me unparalleled elsewhere, unequaled in their poignancy or precision. Shakespeare is worth doing because however difficult his language can be, it is invariably worth the effort simply on a sentence level.

This same sentence-level language is also of a density that breeds alternative readings and approaches, something of crucial value to drama if it is to retain its freshness. The very complexity of the language facilitates, even demands, a certain flexibility on the part of the performers, giving them what directors and actors love best: choices. These choices multiply still further when one bears in mind that theatre is as much about the spaces around the words as about the words themselves, the theatrical product being about material elements not defined or even addressed by the words of the script. Over the generations as the plays are performed and reperformed in wildly different places, times, and cultures, these choices take on immense significance, accruing for the printed word a legacy of interpretive and playable possibility generated from the sheer ubiquity of Shakespeare in the theatre. Staging a line from Shakespeare, then, is to enter into the richest of theatrical dialogues with productions from other places and times, an experience unattainable to the same degree through the work of any other playwright. We stage Shakespeare, therefore, to converse with history and geography, to challenge them even, and thus to define ourselves as part of a larger culture while affirming our own unique position.

Shakespeare's stories have a singular breadth, a scope that ranges from issues of the immediate and personal, through the national or political, to the philosophical and ideological, sometimes in the course of a single scene or character arc. The vastness of the plays is extremely rare, in part at least because the cultural moment in which Shakespeare wrote has passed, and few dramatists today—in tune as they are with their own cultural moment—would attempt to replicate it. The principle holds true for the other innumerable features that distinguish Shakespeare from subsequent drama in terms of the plays' structural dramaturgy, its approach to character, to dialogue, and so forth. To suggest that in quest of immediacy and relevance we should stage only modern plays, therefore, is to limit ourselves to only certain kinds of drama. We stage Shakespeare because doing so gives audiences access to a totally different version of what can be done on stage, and thus a different range of emotional and intellectual experience.

When we stage Shakespeare we necessarily change the play as written both from its status as text and from its original Renaissance performance. In a sense then, we are always telling a different story on the modern stage from that preserved in either of those prior incarnations, but this is no reason to abandon the project. On the contrary, we must embrace these theoretically unavoidable facts and agree to tell stories that are related but different, stories distinctly linked to their textual and theatrical forbears, now rendered new and vibrant through performance. The stories we tell on the stage are linked to Shakespeare by a myriad threads, and though the precise form of the story is different in each production, it remains a negotiation of the original which draws expressly on that original, holding it up to the light, scrutinizing it, turning it so that it refracts the light in new and intriguing ways. Contemporary performance is thus always a response to the play as much as a manifestation of it and, as such, it enacts forms of literary and cultural criticism even as it constructs art and entertainment in its own right. To think that this makes the *Shakespeareaness* of the original irrelevant, a convenience on which to draw while reinventing the story out of whole cloth, is to reduce the complex process of theatrical production to an untenable binary: Shakespeare or Not Shakespeare. Rather, a theatrical engagement with a Shakespeare text is a participation in the construction of meanings, which, as Worthen says, might be called Shakespearean even though they are new and theatrical. We stage Shakespeare to tell stories that speak to us of the past and present by tentative analogy, showing us ourselves and those of other times and places through a kaleidoscopic lens that transforms the original, refashions it so it is both familiar and new at once.

It might also be said that staging Shakespeare provides a unique opportunity for the theatrical exploration of ideas and issues because SHAKESPEARE is more than the sum of its textual parts. SHAKESPEARE is an industry (several, in fact), a culture, a shifting body of thought and values, and audiences therefore come to it in ways quite different from the way they approach the work of any other dramatist. They bring assumptions and expectations that the production can engage usefully, discovering the political in the familiar aesthetics, for example, or the shocking and subversive in what seems safe and reliable. Shakespeare's venerable respectability can be made a stalking horse for intellectual analysis and subversion of all sorts, the plays becoming forum or battleground for larger sociocultural struggles and debates. We do Shakespeare's plays on stage, then, to challenge or explore what SHAKESPEARE—the culture, and all which might be bracketed alongside that culture—assumes or values, using the status of the plays themselves as a way in to the reconsidering of crucial ideas.[1]

The staging of Shakespeare is therefore a unique enterprise, one that holds the text to be of crucial value, while simultaneously embracing the idea that theatre is constructive rather than simply interpretive, local but also global, celebratory and interrogative, even hostile to elements of the play as written or as staged in the past. It is because of Shakespeare's amorphous adaptability—an adaptability made real through the interaction of theatre practitioners with a richly varied and complex text—that the dramaturg's role looms especially large, committed as he or she must be to facilitating a range of playable options rather than insisting on a monolithic SHAKESPEARE whose plays are largely unchanged from one production to the next. Without this commitment to flexibility, to newness and to choice, the question as to why we stage and restage Shakespeare becomes unanswerable, becomes indeed a leaden albatross around the neck of theatrical culture. If Shakespeare on stage is necessarily to be fusty and staid, or is without clarity and immediacy, then we should not do it. We stage Shakespeare because the nature of the plays as the raw material for a theatre that builds as well as deciphers can generate both the familiar and the new, a theatrical and cultural heritage rendered challenging, fresh, electrifying as only good theatre can be.

PART II

PRACTICE

This section of the book attempts to describe the nature of the dramaturg's work in terms of a production's timeline, showing what should be done at each stage and how the dramaturg's shifting function demands not just a different degree of involvement but a different sense of the dramaturgical self. This latter affects thought process and modes of communication as well as the kinds of information the dramaturg brings to the production. Everything I say here is premised on part I, particularly on those larger arguments about the way theatre authorizes itself, rather than being dependent on questionable assumptions about authorship, history, or text. Despite the descriptive tone of this part, it in fact extends and illuminates the arguments I have already stated, attempting to bring the theory into active engagement with the material conditions of theatrical production in terms of praxis. Since the logic and assumptions of the previous part are here invoked (albeit silently) as the guiding principles that structure the dramaturg's methodology, I would urge the reader to frame disagreements with my method not as matters of mere idiosyncrasy (though they may be no more than that) but in terms of the larger intellectual principles that may be at work beneath or alongside them. In such disagreements there can be little room for an unscrutinized "common sense" view of, for example, the sanctity of the Shakespearean text, and while I have no doubt that there are flaws in my attempt to marry theory and practice here, I would refer such disagreements first back to part I. If those disagreements still stand in the light of broader theoretical consideration, then deviation from the practical means I suggest here may well be desirable.

It bears repeating that my sense of the job is very much defined by the conditions at GSF, and though I have found that such conditions are fairly representative of the way other theatres use dramaturgs, there is bound to be a certain amount of subtle variation. In such cases, the application of the theoretical principles already established is doubly important, as is the sense of the dramaturg as a collaborative presence mediating between various communities and modes of discourse in pursuit of an expressly theatrical construct.

SECTION III

BEFORE REHEARSALS

Preliminaries, Casting, and Directorial Vision

Depending on how the company functions and how quickly she was assigned to the project, the dramaturg's work can begin weeks or months before the production goes into rehearsal. The rule of thumb is simple: get in as early as possible. If dramaturgs are kept out of the prerehearsal process, as often happens when they do not press the director and production staff for earlier involvement, or when they get hired merely to handle formal responsibilities such as program notes and audience lectures, they are deprived of the capacity to make some of their most important contributions to the show. They will, moreover, spend the entire rehearsal process playing catch up. A dramaturg who has barely seen the performance script before first read-through with the cast, for example, is at a tremendous disadvantage. The dramaturg may be called upon for advice or insight that she is not ready to give because she is not sufficiently familiar with the text (particularly if the director has made cutting choices already), and she will spend the first vital days of the process trying to get onto the same page as the director, longer if the two have not worked together before. All this is badly wasted time and tends to marginalize the dramaturg right off the bat, rendering her incidental to the show.

The dramaturg who gets involved before the rehearsals begin, however, has the opportunity to share ideas with the designers, to directly contribute to the overall concept of the show and its script, and can otherwise build a working relationship with the director and the rest of the production team, all of which point toward a more fruitful and unified rehearsal experience. Further, the dramaturg who enters rehearsals comfortable with the director and the script and the overall vision of the show is obviously an active and respected member of the company, someone less easily dismissed as the Shakespeare police than one who is flying by the seat of her pants, and still trying to find a niche within the company.

Having said that, some directors and companies assume that dramaturgs don't—even shouldn't—get involved till rehearsals begin, and this will require discussion with both the director and the company manager, since such an attitude is likely to greatly reduce the degree of involvement to which the dramaturg can look forward. The nature of the job needs to be agreed upon (preferably in writing) before any work is done, and I would urge dramaturgs working at all levels to get some kind of contract or letter of agreement in advance of any work on the show. This is not a haggling tool (there is rarely much money available for the adjunct production dramaturg at the small or mid-level company anyway); it is a way of clarifying what is expected of the dramaturg and what the dramaturg, in turn, expects of the company. Such an agreement is important since expectations of dramaturgs can vary and because there is no powerful dramaturgical union (such as actors have in EQUITY) that can meditate disputes (though the some guidelines and recommendations are available through the LMDA [Literary Managers and Dramaturgs of the Americas]). Awkward though it may seem to request a written agreement, most companies are glad of a clearer sense of their relationship with the dramaturg, doubly so if they feel that she is looking for more involvement, not less. A written agreement will help to build a sense of relationship between the dramaturg and the rest of the company, and can help to minimize misunderstanding and conflict about what the dramaturg can and cannot do.

The letter of agreement should state, as precisely as possible, the nature of the dramaturg's role. Companies unused to working with a dramaturg, or whose sense of the nature of that work might need refining or revising, may be glad of the new dramaturg's input on the content of the agreement, in which case it may be helpful to forward to the company's manager or artistic director some suggestions derived from this book, other relevant publications, or practices employed by other comparable companies.

Apart from stating salary or honorarium (if any), the letter should include a list of the dates of any known associated functions or responsibilities in which the dramaturg must participate, such as preshow lectures or talkbacks. Ideally, the letter or contract would also indicate how much attendance in the rehearsals is expected of the dramaturg, expressed either as a percentage of the total rehearsal time, a statement of specific dates or events (first read, dress rehearsal, and so on) at which their presence is required, or (more likely) a general statement about being available to serve as a resource for actors and director.[1] Some letters will suggest that the precise details of the dramaturg/director relationship will be worked out between the two interested parties.

The letter or contract should also address the dramaturg's role outside the rehearsal room, particularly with regard to the editing of the performance script, but also as to whether she will be expected to attend production meetings, and if she should supply the company with some kind of research packet or book in advance of rehearsals. While all this may seem legalistic or unnecessary (particularly if there is little or no money involved), it will give the dramaturg a clearer sense of her position and supplies a baseline to which she can return if problems arise. The LMDA website (www.lmda.org) details several employment contracts according to who is hiring and for what kind of venue. It is unlikely that the dramaturg will be able to press for a larger honorarium until the company has seen the quality and extent of her contribution, and while that might be conveyed by references, program notes, and research packets from prior shows elsewhere, the first-time dramaturg is likely to have to prove herself before she achieves any bargaining power.

The most important person with whom to liaise before the rehearsal process begins is the director, and it is imperative that the dramaturg begin a dialogue with him as quickly as possible. Having already worked out some form of contractual relationship with the company will help to underpin initial conversations with the director, especially if he is unused to this kind of working relationship or is guarding certain dramaturgical functions overzealously. It is important that the dramaturg outlines her sense of what she will be doing to the director right away so that agreement can be reached. If agreement cannot be reached—for example, a director may not like the idea of the dramaturg talking directly to actors, preferring to have all notes go through the director herself—then the dramaturg has to decide how hard to push to get her way. If the director is not in accord with the letter of agreement, the company manager or artistic director may be called upon to politely explain the way the company views the dramaturgical role. If the director remains adamant on key issues, the dramaturg needs to decide how important these areas are to her sense of her job. It is not inappropriate to withdraw from a production if the dramaturg truly feels she cannot work effectively on it, though this should, ideally, be done well in advance of the rehearsal period, and with a clear sense that such an action may well burn bridges. Some directors will be initially wary of giving an unfamiliar dramaturg a lot of access to the actors: if the dramaturg can demonstrate her supportiveness and value in the first days of rehearsal, however, such directorial caution can often be eroded completely. If the director seems to be feeling out the dramaturg in their first meetings (as is entirely appropriate) I would not press hard for things like the right to routinely talk directly to actors, accepting instead that this is a privilege to be earned.

Some directors like to keep the script to themselves before rehearsals begin, using it to test and explore their conceptual approach to the show, and this impulse to privacy can be exacerbated by being presented with a dramaturg who is a "professional" Shakespearean such as an academic. Directors have good reason to be wary of academics and dramaturgs by association, and some are frankly prejudiced against them, believing them to have little real interest in the power, flexibility, and mechanics of practical theatre. This is, hopefully, a misconception that time and a working relationship will destroy, but the dramaturg does not have time if he is to be involved in script editing before rehearsals begin, so he must rely instead on tact. If nothing else, the dramaturg should push for a copy of the script as the director works on it, and should then supply editing notes and suggestions thereafter.

In initial contact with the director, the dramaturg should be attentive, flexible, passionate about the play and about being involved in the production, and very slow to bring up overly scholarly ideas about the text. Some directors are excited to talk theory and politics, but many aren't and can be made wary— even intimidated—by a dramaturg who seems to see the play exclusively in these terms. If these initial conversations take place in writing (many dramaturgs do much of their initial communication by e-mail since the director is often working elsewhere when discussion begins) he should be doubly wary of overly academic language that the director could find off-putting, if not actually baffling. Without being disingenuous the dramaturg should, in his initial dealings with the director, spend more time listening and asking questions than he should making suggestions or preaching the merits of certain readings. The more he seems to view the play solely as a historical, textual, or ideological construct, the harder it will be to win over the director to a sense of a full and open working relationship.

Jane Ann Crum says that some directors want their dramaturgs to be forceful, lest the director find them easy to ignore, and the beginning of the process is the best time to initiate such "aggression" (76). This does not mean the foisting of ideas or material that is otherwise rhetorically or semantically inappropriate. Some dramaturgs, for example, expressly recommend bringing no literary criticism to a director's attention, or suggest bringing no books at all to initial conversations with a director "unless they're picture books" (77). I'm wary of the infantilization such a remark seems to imply, but the point— that the dramaturg should not overly intellectualize the process at least until he has a sense of how comfortable the director is with such things—is a good one. This does not mean that the *content* of literary criticism should not find its way into those initial conversations. On the contrary, the dramaturg should be well-read and armed with interestingly playable ideas derived from literary criticism, but shaped and clarified in ways the director will find useful, not overly abstract, overly historicist, or patronizing. Such criticism—or at least the ideas it deals with—is a vital part of these exploratory discussions, the dramaturg standing in for the academic community as part of a collaborative and educational exchange with the theatrical world figured by the director. Professional literary studies are an excellent locus for the richest, most complex, and subtle analysis of the plays and their various contexts, and though some theatre people can be dismissive of their value (even their validity), such attitudes should not taint the dramaturg's work. To ignore literary criticism in preparing for a show is to reduce the range of intellectual options that can be taken to the director, whether those concerns be formal, aesthetic, or sociopolitical. Monographs or books of critical essays on a particular play can be useful depending on their approach, as are individual pieces in literary journals such as *Shakespeare Quarterly, Shakespeare Bulletin, Shakespeare Survey, Medieval and Renaissance Drama in England, Cahiers Elisabethains, English Literary Renaissance, Renaissance Drama,* or less period-specific journals such as *English Literary History* (*ELH*). All can be searched through the MLA's online database available at academic libraries. Such material is, of course, aimed at the professional scholar, and while it may be invaluable to a dramaturg's sense of the discourse around the play, it should rarely be presented to actors or directors unmediated.

More directly shareable might be studies of a play's performance history. Arden have a series of books detailing the performance history of the plays at the RSC, and both Manchester and Fairleigh Dickinson University Presses have excellent book series, the former focusing largely on the play's twentieth-century performance history, the latter emphasizing the entire stage history since the Renaissance, and these can also contain excellent insights into the play both as text and as performance.[2] More general books on Shakespeare's plays in performance (see Kennedy and Bate and Jackson) are listed in the works cited and appendix sections. Similarly, handbooks or companions to Shakespeare aimed at undergraduates such as that by Russ McDonald contain valuable (and readable) contextual information that the dramaturg can share with the director. Similarly, historical research may prove extremely revelatory, be it of Shakespeare's own period, the period in which he set his plays, or the period in which the *production* will be set, all of which may be quite different. Historical consideration of Shakespeare's own theatrical conditions may also be useful, and though I would avoid the use of such material for merely archaeological purposes, it can be constructive to see how certain scenes might have been staged (see Dessen), what kinds of doubling might have been employed and who was in the audience (see Gurr, *Playgoing* and *The Shakespearean Stage*), how Shakespeare's characters were linked to figures like the Vice in earlier drama (see Weimann, *Popular*), or how audiences understood the way Renaissance actors presented character (Dawson, "Performance and Participation").

Nevertheless, some directors (and actors) sometimes become defensive at the idea of any significantly intellectual component being suggested by the dramaturg, and it helps to remember David Copelin's response to the resultant skepticism or hostility that the dramaturg can sometimes face when he or she pushes for ideas:

> No, [. . .] dramaturgs don't want to replace human emotions with abstract ideas, and we certainly don't accept the common error that human emotions are necessarily "warm" while abstract ideas, or any ideas, are necessarily "cold." We know how many crimes have been committed under that phony standard. We'd rather recognize the joy of ideas as embodied in and by works of dramatic art, and their concurrence and compatibility with emotion. We're not afraid to have ideas if they are presented in truly dramatic form (not like raisins dropped into oatmeal). Patronizing an audience, assuming its stupidity or lack of interest in ideas, is the worst kind of elitism. Such elitism is not just anti-intellectual, it's inhumane. It's poison. (Copelin 20)

Though "intellectual" is sometimes brandished as a dirty word in theatre, ideas are worth fighting for. The dramaturg may have to find other ways to bring the issue up than bombarding the director with historical tracts or theoretical monographs, but this does not mean that the ideas themselves have to go; with the right packaging most directors will be as excited by them as by a production's emotional or aesthetic choices, particularly if they are an organic part of the whole, not, as Copelin says, merely "raisins dropped into oatmeal." That said, the dramaturg is not the director and thus needs to be aware of the fact that as such he "suggests his opinion to the director but does not force it on him, and understands that the final decision in all matters raised in discussion is necessarily the director's" (Katz 15).

The dramaturg, as I have said, serves the show, not the text or his own intellectual agenda, and much of that show will be sketched out by the director. Everything the dramaturg does, therefore, must be accomplished with a clear sense of what the director is thinking about or intending for the show. A dramaturg working on two different productions of *Much Ado*, for example, would produce two different scripts, because the directors are different and the shows will have a different sense of themselves. The very best thing that a dramaturg can do, then, in his first dealings with the director is to ask the big questions that will lead to a sense of what the director sees coming from the production: what first drew you to this play? What do you see as the core ideas, relationships, or images? How do you envisage the journeys of the key characters and how do you expect the audience to feel or think in the course of these journeys? What design elements do you expect to use? How do you see the world of the play, its structures, colors, and energies? What do you ultimately see the show being *about*?

If these questions are asked weeks or months before the rehearsal process begins, the dramaturg should expect their answers to be tentative and provisional, but some answers (however much they might change or be scrapped altogether in the course of the process) are better than none since they will reveal at least the general ways in which the director thinks and how he or she is starting to approach this particular play. Even provisional answers to these questions will help the dramaturg in his early work on the script and in writing program notes (which are often due—particularly in a repertory company that produces a single program for several shows—before rehearsals begin).

The least helpful answers are, of course, versions of "I don't know yet. We'll see what happens when we get into rehearsal." Such a wholly organic approach is perfectly valid, even desirable (though statements professing absolutely no preconceptions about the show are rarely strictly accurate since the director will have some working assumptions), but the success of such a "blank slate" approach is also dependent on the length of the rehearsal period. If the show will have three months of rehearsal or more (three months in which the actors are not focusing largely on other productions), such an approach requires that the dramaturg wait on the script until things have started to come into focus, and while this waiting period can be a little nerve wracking it merely requires that the dramaturg reset her schedule. It does not, of course, preclude the drafting of preliminary notes for textual cuts and modifications.

If the show has a less luxurious rehearsal period, however, as is the case for most companies in the United States, this wholly organic approach can prove difficult for the dramaturg's prerehearsal work. In such a situation, I would suggest a conversation with the director in which the dramaturg press delicately for a general sense of the show (something directors will often reveal in their assumptions even if they don't want to commit to key issues or choices yet) framed as an interested attempt to ensure such a vision is represented in the script and in peripherals such as program notes. If the director still cannot produce enough to provide a starting point, it may be worth asking for advice from the company manager who may, in turn, pass along a sense of urgency to the director. In no case should the dramaturg dismiss the director's "organic" approach, though it might be helpful (particularly if time is short) to suggest that the actors and the design team would benefit from a limited number of decisions or questions being made in advance. Again, the nature of these questions might be useful to the dramaturg's sense of the show. If more is still required, it might be the time to tentatively raise literary readings or scholarly debates about the play, floating them to see what the director leans to or away from. The director may respond positively to suggested (or forwarded) reading, either about critical approaches to the play or accounts of previous (or analogous) productions. Again, what the dramaturg is hoping to get from such debate is less a definitive statement on any given issue than it is a feel for the director's impulses and preconceptions. If nothing useful is forthcoming, the dramaturg should begin the work anyway, keeping decisions speculative and provisional, talking to the director constantly throughout the process and sharing any thoughts he has along the way.

While it might not be the best strategy to overpower the director with theoretical or critical works, some form of carefully tailored material (something, perhaps, that might later be adapted into a handout for the cast) might help to generate conversation and ideas. Mark Bly (dramaturg/literary manager at the Guthrie since 1981) says he prepares a document containing an extensive production history full of visual images, initially very disparate (even contradictory) but tailored to what he thinks will interest the director (Moore 108–9). Such an approach creates a nonconfrontational and discursive mood at the outset as well as—hopefully—generating or revealing real ideas.

One way that the dramaturg can learn about the director's leanings on the show is by paying attention to the process and outcome of casting. Many dramaturgs think casting to be outside their domain, but if we consider the larger intellectual concerns and structures of the production to be, in part, the realm of the dramaturg, then casting can benefit from the dramaturg's input since it significantly shapes the way the show will produce meaning. If the dramaturg is permitted into auditions as an observer or sounding board, he can glimpse the director's sense of the show as she currently imagines it through what she asks of the actors and what she values in their work. While casting is properly a directorial concern, the dramaturg can also be of real assistance in several other ways.

First, the dramaturg can help the director think about the basic issue of how many people to cast by scrutinizing the play for options where certain minor roles could be cut, or rolled into one, and where doubling possibilities will not produce logistical problems (moments where two characters appear on stage together, for example, or where an actor playing two roles has insufficient time to change costume). This is a simple, mechanical task that requires a thorough charting of the characters in the play on a scene-by-scene basis, similar to that which will be produced by the stage manager when rehearsals begin. Most directors will do this themselves, but the help of the

Figure 10.1 *A Midsummer Night's Dream.* GSF 2000, Dir. Garner: Mark Kincaid (Oberon), Saxon Palmer (Puck), Janice Akers (Titania), and Thalia Bauldin (First Fairy). Photo: Kim Kenney.

dramaturg can make the task a good deal less arduous. Furthermore, through recourse to reviews of other productions, the dramaturg can raise matters of how doubling has been used in the play before and to what effect. What, for example, are the implications of doubling Oberon with Theseus and Titania with Hippolita in *A Midsummer Night's Dream*? How does such a casting decision alter the dynamics between the characters or the politics of power (particularly in gendered terms)? What is gained and lost when a single actor doubles as both Antipholi in *The Comedy of Errors*, as Viola and Sebastian in *Twelfth Night*, or as Hermione and Perdita in *The Winter's Tale*? In less literal doubling (what Alan Dessen calls "conceptual casting" in which the audience makes significant links between characters who they recognize to be played by the same actor) what is conveyed by doubling the three murderers of Banquo with the weird sisters in *Macbeth* (or Duncan with the Porter or Old Siward), or Posthumous with Cloten in *Cymbeline*? Most of Shakespeare's plays contain the possibility of suggestive or thematically evocative doubling and discussion of such options must occur before casting takes place. If the dramaturg raises such an idea afterward, the director may well feel that a significant opportunity was missed. Communal work of this kind early on further develops the sense of a relationship between director and dramaturg, facilitating dialogue, and, ultimately, building a sense of shared ideas and investment in the project.

Second, casting is an excellent point to begin work as that crucial dramaturgical function, the sounding board. Directors often have no one to speak to during casting except for the company manager or artistic director who has

Figure 10.2 *The Winter's Tale.* GSF 2001, Dir. Garner: Rob Cleveland (Polixenes), Anthony Irons (Florized), Allen O'Reilly (Leontes), Tim McDonough (Camillo), Janice Akers (Paulina), Carolyn Cook (Hermione), and Jessica Andary (Perdita). Photo: Rob Dillard.

to pay attention to matters of pay scale and budget during casting or who, in the case of a repertory company, may be primarily interested in assembling a company that will work for other directors and shows. It can thus be invaluable to have the production dramaturg—someone invested in this show only—there to share thoughts on the different perspectives and energies that certain actors might bring to a role. Again, such discussion can only help to clarify a sense of what the director envisages the show as being, something the dramaturg will find invaluable when left alone to work on the script. Moreover, if the dramaturg is visibly involved in such things from the outset, the other members of the company (including the actors who will remember the dramaturg from auditions) are more likely to see him as an intrinsic element of the show, someone with the ear of the director, someone to be trusted, all of which will increase the dramaturg's possible contributions thereafter.

As someone invested in the intellectual cohesion of the production, the dramaturg can also be a useful resource during casting when dealing with anything that could have an impact on what a production is perceived to mean in areas that are potentially controversial or politically provocative. Casting involving race and gender can be particularly tricky, whether the issue concerns the casting of an actor whose ethnicity suits the role (a black Othello, a Jewish Shylock) or when the role is being cross-cast (a character originally written as white or male being played by someone of color or female, for example). For such situations, some general principles might be of use.

The dramaturg's first function in such instances is to open up interesting and productive theatrical choices while ensuring that the director is aware of possible controversy, even if that controversy has as much to do with the dynamics of the local community as it does with anything inherent to the text or the director's approach to it. In such cases, the dramaturg's role is not to insist on a particular agenda but to ensure the director is clear on the possibilities and how they might be read. There is no question that both *Othello* and *The Merchant of Venice* are built in part on racism, though the extent to which the production itself is incriminated by that racism (rather than exposing it) is a matter of execution.[3] It is also clear that staging a play that is about racism but not itself racist does not necessarily require turning an Othello or a Shylock into flawless heroes. The dramaturg's job in these matters is to ensure that these concerns are met head on and tackled, not brushed away as too awkward, difficult, or irrelevant to merit consideration. However much a director may pronounce Shylock's Jewishness or Othello's blackness essentially irrelevant to the show, it will not be perceived as such by the audience. For better and/or worse, race does signify on the stage, however much the production may wish to side-step it with the rhetoric of a more general humanism.

For this same reason, racial cross-casting has become a powerful and divisive issue. While many people want their Shakespeare on stage to be democratically inclusive, there are reasons to be suspicious of wholly color-blind casting, not least because race is usually inscribed in the body of the actor in ways the audience *sees*, ways that therefore alter the construction of meaning. To cast a white Ophelia and a black Laertes, for instance, does not simply create a sense of visual paradox for an audience looking to see a brother/sister relationship that is somehow without race, it also alters the ways the roles will be perceived of themselves. This might be a good thing, or it might not, but it needs to be anticipated. In a recent GSF production of *The Winter's Tale*, the actors playing Polixenes and his son Florizel were both African American, a decision that not only made perfect internal sense, but that added an extra spur to Leontes's sexual jealousy.

We live in a racially aware culture and that awareness—even if it is relatively unencumbered with prejudice or stereotypical association—should be expected to signify on stage as it does off. Where a black actor plays a servant whose race is unspecified in the text, he *may* be seen by the audience as a slave. Whether this is a good or a bad thing for the show will vary, of course; the important thing is that the possibility is considered in advance, and angry letters from local community leaders do not catch the company completely off guard.

A similar—if less highly charged—debate may be triggered by cross-casting in terms of gender, giving, for example, female actors male roles (whether they play them as men or as women). In such cases, of course, it is their contemporary implications that take precedence over historical concerns tied to the use of cross-dressing on the Renaissance stage (though redressing that particular balance can be a serviceable rationale for cross-casting in the present).[4]

Cross-casting is paradoxically more difficult if limited only to a few roles; productions that reverse *all* the genders or races of the characters, or play with an *all* black or *all* female cast, tend to carve out a clearer attitude to the core issues that underlie the play. Of course, such productions are heavily concept-driven and are not typical of Shakespeare on stage, even if they are much more frequent than they used to be. Dramaturgs working on productions that are less sweeping in their cross-casting choices should always keep the basic idea—that gender doesn't simply disappear on stage—uppermost during casting. Even when a woman plays a man's role as a man, her own gender will almost certainly be visible (at least intermittently) to the audience in ways that alter their sense of what this particular casting choice—and thus the show in general—means.

Cross-casting will, in some quarters, raise the issue of authenticity, which I addressed in part I, and the dramaturg should be prepared to meet such challenges with informed and reasoned debate. Indeed, it is not uncommon for some of the questionable rhetoric about Shakespeare's intent—or of adhering to a "purist" notion of the text in such matters—to come from within the company, even from the director. As ever, such things have to be handled diplomatically, but if the dramaturg can tactfully correct mistaken assumptions, the range of possibilities for the show will almost certainly increase. If a close eye is kept on the theatrical modes of signification released by issues such as cross-casting (and this includes the fact that some people will, however illogically, view a cross-cast production as problematically overadaptive) the show will, at least, remain authentic to its own rationale and logistics. As the dramaturg is tied to the show's intellectual coherence, so must she be committed to making the show intellectually provocative, adventurous, and powerful, all of which can be initiated by bold casting choices.

As with involvement in casting, whether or not the dramaturg will be kept in the loop as far as initial design concepts are concerned will vary from production to production and company to company. As a general rule, I think that the dramaturg's direct connection with the designers should be occasional, and either in the company of the director or as part of communication expressly authorized by the director, such as the passing on of research to the designer team. This more restrained interaction stems partly from the fact that most dramaturgs' primary area of expertise is the construction of intellectual meaning through language, and partly because the work of a design team takes place largely out of the gaze of the director who tends, having made suggestions, to see more product than process. Unlike working with actors whose performance develops incrementally under the director's gaze and can thus be shaped on a moment-to-moment basis, a director's awareness of a design element such as costume tends to move through distinct steps. If a dramaturg misspeaks while talking to an actor, then, or suggests an avenue of approach that the director decides not to pursue, the error becomes instantly apparent in the rehearsal and can be addressed. If the dramaturg misspeaks to a designer, however, it may be some time before the extent of the miscommunication becomes clear to the director, by which time a good deal of energy and resources may have been wasted. As no designer wants to dismantle elements of costume, set or props built on the faulty advice of the dramaturg, no dramaturg wants to be left looking like he or she has made the director's job harder rather than easier. That said, since all the semiotic elements of the performance (visual and aural) have some hand in the construction of those meanings, it is essential that the dramaturg understands the production's nonlinguistic dimension and is kept aware of what choices the design team are making, so that he can factor in such elements to his sense of the show in general and the script in particular.

It should be said that while the good dramaturg is generally encouraging and supportive of the other members of the company, there are times when he should be prepared to raise serious objections to preliminary concepts if he thinks they are inappropriate or unthought-through. Such objections should first be raised as questions: how does this set, for example, connect to the story or its world, how can it be made to connect better, and if it doesn't and won't, is there a useful statement there, or should we reconsider the concept? This is not to say that an Old West *As You Like It* should be dismissed out of hand. Rather, it means that the concept should be explored for its capacity to frame or shape the story, its ideas and relationships, and if the dramaturg thinks that it feels gimmicky or irrelevant, he should say so, tactfully, to the director (not the designer directly) before undue work goes into the project. It could be that the concept simply needs a little rethinking, tying better to the play, but it may also be that the concept will always feel flawed, "painted on." While it is not the dramaturg's call to determine whether or not the concept stays, it is his duty to raise doubts about it before it is too late. To acquiesce to something one believes will not work is to share responsibility for its eventual failure.

11

THINKING ABOUT SCRIPT

Preparing, or helping to prepare, the script for the production is possibly the most important work the dramaturg can do. It is where his expertise is of most material value and it is one of the single most important factors in laying the groundwork both for effective rehearsal and for a clear and provocative show. The dramaturg should thus do everything in his power to ensure that he has a significant role in the cutting and editing of the script. Indeed, I would go so far as to say that the dramaturg who is not involved in this crucial task is really doing less than half of his job, and he is almost inevitably going to be peripheral to the show. The dramaturg who only gets involved with a production after the script has been compiled has forfeited much of his opportunity to shape the show in the medium that is generally supposed to be his area of specialty: the words of the play.

The idea that Shakespeare needs no editing for performance, presumably because of the genius of his writing, is flawed in many ways, some theoretical (the elusive pursuit of a textual "purity," for instance) and some practical. Having already addressed the theoretical problems with such an approach in part I, I will here merely state what is obvious about the practical problems of doing Shakespeare "uncut." Before I do so, however, let me say that the deep unease some directors and critics feel about editing the Shakespearean text at all is an exclusively twentieth-century construct. From the Restoration well into the nineteenth-century, when Shakespeare was performed it was usually in a heavily adapted state, the Renaissance original being shaped to the tastes and ideas of the current period in ways that frequently left fewer lines by Shakespeare than by his latter day "collaborators" in the final performance script. Thomas Sheridan's highly influential 1752 production of *Coriolanus*, which ran fairly consistently in London for twenty years, used less than half of Shakespeare's original lines, added almost as many by James Thomson and another seventy-one by Sheridan himself. In doing so, Sheridan enacted a common attitude of his day, prompting one reviewer to declare, "the gross Absurdities of Shakespeare in Point of Time and Place, have in a great Measure been amended . . ." (Ripley 108).

Plays such as *King Lear* were completely restructured, their endings altered, their characters extensively rewritten, so that accounts of the plays in performance frequently bear little resemblance to the play as we currently know it. The theatre managers and critics of the day were a good deal less overawed by a sense of Shakespeare's genius than in more recent years, when Shakespeare's place in educational curricula and the larger culture in general has rendered him not so much a playwright as an icon, a symbol of something beyond mere theatre. While the twentieth-century "rediscovery" of a less heavily adapted Shakespeare by directors like William Poel is generally considered a step forward, there is still something to be learned from the Sheridans and Keanes of the theatrical world, whose commitment was to the stage as they understood it in their own day.

In the context of a Nahum Tate *King Lear* (in which Cordelia lives and marries Edgar, the only version of the play seen on the professional stage for a century and a half), what I am to advocate in terms of editing seems positively tame, even Bardolotrous, but since it will still seem radical to some, let me quickly recap some of the reasons why the dramaturg *must* edit Shakespeare for performance. First, many of the plays are simply too long for conventional production today (and probably were too long as printed for the Renaissance stage too) and thus require trimming to bring the show in under two and a half hours. The plays were written for characters whose thought is in the words or on them, not between them as tends to be the case in more modern plays, and maintaining an uncut text can thus aggravate the problems of length, either by having the actors deliver lengthy exchanges that, for a modern audience, turn all too easily into walls of sound, or by leading them to act in the silences, lengthening the play still further.

Second, the plays are the products of their period and are thus littered with archaism and contain, even where the actual diction is familiar, a poetic density that can prove baffling to an audience that no longer hears the language as the playwright's contemporaries did. The plays are shot through with topical references to incidents, places, bodies of thought or belief that, though still potentially quite clear in their wording, remain opaque or lacking in significant nuance for the modern audience, which does not grasp the original association. Finally, strict adherence to the text limits the range of interpretive possibilities in ways at odds with both what we know of how those texts came into being and were originally used, and with the transformatively and constructively creative energies of theatre.

All those things said, directors and dramaturgs betray a curious shame-facedness about cutting Shakespeare, and Sidney Homan is representative of directors who cut the plays in the name of practicality (usually merely to abbreviate running time) but talk about such cuts with embarrassment and, frequently, regret.[1] Some of this embarrassment is surely disingenuous—the placating of an audience assumed to be Bardolotrous—but some of it is (as seems the case with Homan) obviously genuine, and suggests a real anxiety about tinkering with the sacrosanct text. Sometimes this anxiety stems from the belief (to my mind problematic and historically dubious) that the text contains specifically theatrical prompts for the actors, which comprise a kind of map for characterization and scenic rhythm built into the wording, meter, and punctuation of the text (often, more problematically still, the First Folio).

Sometimes the anxiety about cutting is less specific and thus harder to explain, though it seems to inhere in the author's genius and or cultural status, factors that make editing his work tantamount to blasphemy. All historicist and theoretical arguments aside, it seems particularly strange that theatre practitioners should feel such subservience to a notion of Shakespeare that is not merely conservative but textual to a degree which utterly marginalizes the work and authority of the theatre itself.

Editing the script is not merely a necessary evil executed to pander to audiences who (the embarrassed editor seems to suggest) don't have the attention span of their Renaissance forbears. It is an essential part of the transition from the play as a specifically textual entity into script: part of the raw material of performance. As such, the editing process enacts the play's shift in genre as it moves from page to stage, and the words are thus neither the origin of all that follows in the production nor the ultimate purpose of that production (despite the dubious tendency of some dramaturgs to think of the production ultimately as a vehicle to learn about, or worse, teach about, the text). The words of the text—yes, even Shakespeare's words—must ultimately serve the production and the performative moment. The editing process then, far from being the unholy stealing from the temple that some critics and directors assume, is a crucial element in turning the book into theatre.

And that notion of "pandering" to an audience needs pretty serious scrutinizing too. It's an appallingly smug and condescending attitude, particularly when one considers that few of the people who say such things agree about what makes the text so rich or complex in the first place. While it is certainly possible to oversimplify a script in the editing by stripping it of its aesthetic or conceptual density, or by otherwise executing the editing *badly*, this in no way undermines the logic of script editing of itself. Theatre presumes (and needs) an audience, and audiences are made up of real people from real communities. If a production uses a script that the company knows its audience simply will not understand, what exactly are they trying to do? To overshoot an audience in the vague interest of being "true" to the text is to turn the communicative aspect of theatre on its head and to privilege textual authority over the collaborative core of the theatrical experience. Since the audience is central to the ways theatrical meaning is created, a script that does not take that audience into account will produce something that is both patronizing and pointless. Even the most educationally minded dramaturg must recognize that for an audience to be stimulated by a show—to learn from it, even—they have to be able to understand it.

This does not, again, mean a "dumbing down" of the script. It means escaping from a sense that the show's authority is predicated on a species of literalism, on its adherence to Shakespeare's high cultural and textual traces. It means foregrounding the voice over the printed word, and using a specific sense of who will hear those words to shape their communicative imminence. This does not mean rewriting the play in a modern idiom, or peppering it with contemporary references (though those things are possible within a suitably adaptive approach to the play), but it does mean a conscious addressing of *everything* in the script that, in the heady air of performance, is unlikely to generate productive meaning in the minds of the expected audience.

That last reference to an "expected" audience, though fraught with tentative caution, is important. A university production, after all, playing mainly to students and professors of Shakespeare has to approach the script differently from a production touring inner city high schools where Shakespeare is sometimes close to being a foreign language. Recognizing and incorporating this basic truth into the editorial process is part of the collaborative construction of meaning between company and audience, a function the dramaturg emblematizes, and I would go so far as to suggest that any dramaturg who refuses to make such concessions to audience should probably not be doing theatre (or teaching) at all.

There are many ways to approach the editing of a performance script. Sometimes the dramaturg may be asked to excise entire scenes, major characters, or plot points in order to focus attention on the concerns the director wants to throw into sharp relief. While this is a legitimate approach to extensive adaptation, cutting of this kind is sometimes used merely as a way of streamlining a play without doing the more detailed work of thinning the entire text. When such a method is used without a clear agenda in mind, the production often suffers because the play's scope is reduced and—more importantly—because such a limited approach to editing often leaves the remaining material largely untouched, and therefore still dotted with moments that are baffling or arcane. Indeed, I think that cutting in large chunks (a scene here, a character there) is often done—like the sudden removal of a Band-Aid—because the company thinks that more detail-oriented cutting is not just hard, it's uncomfortable, even painful, for all the reasons mentioned above. Take out an entire subplot, such an approach seems to say, and the play gets nicely shorter while leaving all that wonderfully unaltered *Shakespeare*, but change words at the sentence level to make the utterance clear and the result is somehow no longer *Shakespeare*.

Such an approach is highly questionable, rooted as it is in an unholy alliance of laziness and Bardolotry. It results in the fragmentation and impoverishment of the play while still incorporating material that needs footnotes to be comprehensible. I argue, therefore, that the dramaturg should proceed on the understanding that, unless the director actively wants to remove large, intact elements of the play as part of the stage adaptation, editing should be done systematically *throughout* the text. In some cases, this judicious "cleaning up" of the script will reveal the value of keeping elements the director might otherwise have been ready to lose.

Before the dramaturg begins work on editing the text, transforming it into a performance script, she should do all she can to get answers from the director to the questions I raised in the previous section. Time spent editing the script without a sense of the director's vision of the play in mind can easily become time wasted. If the director has found, for example, a particular set of recurrent images significant (images that may be echoed in parts of the show's visual design), the dramaturg needs to know so that she will not cut them, or so she can clarify them, by pruning around them or making the words of the images more comprehensible to the audience. The presence of minor servants on stage can reinforce an impression of hierarchy that a director might think is important to his sense of a scene or character but which a dramaturg might be tempted to cut or roll into one; consider, for example, the number of servants surrounding Orsino in *Twelfth Night*, and how the impression of all those vividly present bodies being supplanted by Viola/Caesario in only three days resonates. Knowing the basic layout of the set (how many entrances will there be? Will there be an upstairs area or a place to hide?) will help the dramaturg to view the script in the pragmatic terms of actor movement and logistics. In short, the more information that the dramaturg has about the director's sense of the show before she begins to work on the script, the more profitable her work is likely to be.

Above all, the dramaturg and director should come to some agreement about the extent of the production's adaptive strategies, particularly with regard to the script. Should it feel and sound largely uncut or does it have the range to make contemporary references? Will the original songs be kept, or will something different (more modern) be substituted and, if so, what kind of substitution would that be? If the setting of the show means that there will be a disjunction between the word and what the audience sees (the character talks about his "rapier" but he's actually going to be armed with a pistol), should the words be changed and how conspicuously? More generally, should alterations to the language announce themselves to the audience's ear, or should they attempt to pass unnoticed? What are the director's priorities where cutting is concerned: clarity, brevity, naturalism? How does the director weigh the poetics of the language, digressive allusion, the maintenance of its metrical cohesion, or extended rhetorical tropes? How important is the play's Classicism or its use of other stories and ideas that the modern audience probably will not know?

The director may not be able to answer all these questions and, even if he can, such answers do not automatically trump (at this speculative stage) the dramaturg's own instinct on such matters. Moreover, what seems important at the beginning of the process often changes in rehearsal. For example, many directors may express initial disinterest in maintaining iambic pentameter wherever it appears in the play, but if the dramaturg's cutting butchers the rhythm of the line, he—or the actors—may see things differently when those lines have to be spoken aloud. The director's answers to these text-specific questions (like those about larger issues of concept) should thus be taken as a general guide, something to keep in mind during the work of editing, rather than a hard and fast rule that strips the dramaturg of judgment. On the contrary, the dramaturg is there to lend a particular expertise, and most directors want to see that expertise come through in the editing process, even if it is contrary to some of the director's own initial impulses. It is better to make tentative editorial choices and flag them for the director's consideration, than not to make such suggestions for fear of deviating from the director's general principles.

Above all, the dramaturg should remember that, as with all things in theatre, the editing of the script is provisional and collaborative. The script will not (and should not) be finalized in a single pass by the dramaturg. Editing a performance script is a process, one that requires feedback from the director at regular stages and which continues in the rehearsal room. The effective dramaturg will always view the editing of the script (particularly in its initial stages) as the construction of informed and reasoned suggestions. The director will not always agree, and here, as elsewhere, the dramaturg must carefully consider the implications and impact of the disagreement for the show as a whole, picking her battles as she goes.

The next stage is to select a text or texts on which to base the performance script. There are numerous different Shakespeare editions, published both as individual volumes and as collections or complete works, and the choice of the most "authoritative" text is difficult because many of them are different in content in ways that reflect a difference in attitude, rather than in matters of "accuracy." What should be generally avoided are editions whose text was prepared a long time ago and are thus outmoded in editorial thinking and method. It was not uncommon, for example, for nineteenth-century editions to make extensive changes to the text (sometimes censoring language deemed unseemly) without announcing those changes to the reader, to pursue other erratic editorial practices, and to include few or no clarifying footnotes. There are many cheap editions (particularly complete works collections) published today by printers who keep their costs down by reproducing a text that is no longer in copyright and thus demands no payment to an editor. Though the edition looks new, the text was prepared around the turn of the last century. If in doubt, check the copyright date not the publication date. Anything predating 1940 is unlikely to be of much value.

In general, it is best to have at least two "good" modern editions, and it is often useful to deliberately choose those that are most different from each other while still being authoritative. The Arden third series is generally considered the most exhaustive in scholarly terms (though some theatre practitioners find their extensive footnotes off-putting), though it is not yet complete, and many of the second series are now rather dated. The Oxford (the basis of the Norton complete) is probably the most controversial series and uses a deliberately historicist approach, often reverting to versions of the plays that the editors claim are more in keeping with the way they first appeared on stage and in print, before centuries of textual editing. The Oxford thus prints two versions of *King Lear* (the Folio and the Quarto), calls Cymbeline's daughter not Imogen, as in the First Folio, but Innogen (based on an early account of the play on stage, and on the name's appearance in source material and other plays), and renames Falstaff "Sir John Oldcastle," the name the character originally had before the company was pressured by the remnants of the Oldcastle family to change it. The new Cambridge series, by contrast, tends to give greater weight to the play's recent performance history. Other collected editions such as *The Riverside* or that edited by David Bevington (which provides the basis for the Folger Shakespeare Library individual editions and for the texts of the Bedford/St. Martin's *Texts and Contexts* series) are also extremely solid. All of these will provide an excellent source of textual variations and introductory materials. Facsimiles of early texts such as the First Folio can be of good secondary use, though the above editions will cover the source of their choices in their textual notes.

Some companies will buy a single volume edition of the play for the actors to use, but I would argue that a specially printed version of a text already edited by the dramaturg and director is generally superior. Extensive editing (particularly changes to individual words) quickly fill the space in a single volume, space the actors will want to use to make blocking or character notes, and takes time to replicate in each copy. A more effective and cleaner alternative is to present the actors with a letter or legal-sized binder (A4 or Foolscap) containing a script typed up on one side of the page only (so that the actors and stage manager have a full blank page next to each page of script on which to make notes) specifically for the company's use. There are several on-line versions of Shakespeare's plays available on the web, which can be copied and pasted into word processing files, thus serving as the basis of the script.[2] While electronic versions mean that the entire script does not have to be typed out, they will need to be studied extremely carefully and with reference to the more authoritative print editions mentioned above as they are often erratic and dotted with errors. The HTML format can also provide occasional difficulties in adjusting line spacing and indentation (usually fixable by reformatting the text using the "style" box in your word processing software), but they are generally a good laborsaving starting point for the editorial process. A downloaded text is useful because it can be revised and modified repeatedly and easily, the editor storing different versions as she works so she can check or revert to earlier passes, and because the editor can write in notes, rationales, or suggestions to herself or the director without cluttering the text unduly. The editor can, furthermore, use font size, bold, or italic script, and combinations of symbols or brackets to highlight sections the editor has modified in each new pass so that they stand out.

12

Preparing the Script

As elsewhere in this book, the methods I propose here should be considered a guide or set of principles, rather than an absolute methodology, though the editorial sequence derived from those principles is one I have found to work effectively since it gives distinct phases to the project. I recommend beginning with an attempt to render the script clear, more major cutting being reserved for later when the text has been made more playable and its remaining difficulties or problems are thus more apparent. Once this clarification process has been completed, a second pass that handles the more extensive cuts or changes (within the adaptive parameters already agreed to with the director), should be completed. I recommend three complete passes on the script before the first read-through with the cast—the third being a detail-oriented polishing—by that time the script should look finished, even if some details are still under discussion. The exception to this rule is when the director is committed to an exhaustively exploratory rehearsal process and is beginning with very few assumptions about what the resultant production will be. Even then, however, since most companies do not have the time to explore *all* textual options in rehearsal—and since even companies that do are unlikely to want to burden the actors with every textual variant or possible choice—I strongly suggest going into rehearsals with the bulk of the textual editing already completed, on the understanding that some (hopefully small) changes will become necessary or desirable during the process thereafter.

The script whose construction I describe in this chapter takes a middle ground, neither "purist" in its attitude to the textual original, nor heavily adaptive. I offer this not as the only—or even necessarily the best—model, but as that which is most common, that which most audiences and acting companies are best prepared for, and one that remains intellectually defensible while still claiming to be both "Shakespeare" and the stuff of good, practical theatre: a staging of the *work* in the senses I outlined in chapter 7. Taking this middle ground is also a practical matter for a book such as this, since more adaptive strategies are too varied for extensive consideration, and even with world enough and time I would hate to reduce their creative energies to mere enumeration. What I offer here might be considered, then, a baseline, a point from which more adaptive productions may begin, insofar as they are using substantial pieces of the Shakespearean original in their performance script. The more conservative "purist" position concerns me less for reasons I have already discussed in part I, though it is certainly possible that some dramaturgs will have good reasons based, perhaps, on a uniquely academic or otherwise unusually textually invested audience, for wanting to maintain a script that is closer to a textual original (a production of *Hamlet* Q1, for example), than that which I advocate here. While circumstances might call for a more conservative approach to the script, the dramaturg should not lose sight of the fact that much of the "purist" position remains highly questionable of itself, even if the production does not have to worry about making the text shorter or more comprehensible. Even small university productions attended exclusively by literature professors and graduate students need to treat the theatre as a theatre, not as a vehicle merely for the exercising of a critical faculty upon an entity still considered basically textual. Indeed, for all their investment in the letter of the text, literary critics should be more— not less—aware of the problems with treating theatre as a mere conduit for the book.

The first step in script editing is to know the text as well as possible and this, naturally, means reading the play several times. I would also make a point of getting to know the textual history of the play by looking over the introductions and textual notes in several of the modern editions I mentioned above. These can be extremely revealing and will draw attention to discrepancies in the earliest printings that the dramaturg might not otherwise perceive, but may prove useful. In *Hamlet*, for example, there are a number of well-known textual cruxes dependent on different versions of early texts. Hamlet's first soliloquy begins in many editions with the lines, "O that this too too sullied flesh would melt / Thaw, and resolve itself into a dew!" (1.2. 129–30). The term "sullied" is a modern editorial choice suggesting "defiled." The Second Quarto's choice (1604) is "sallied," which is often taken for a misprint, but could suggest "besieged" (as in to "sally forth" into battle). I doubt the value of the Second Quarto's version for a performance script today, "sallied" being somewhat obscure, though it may serve as a useful note to the actor to clarify or deepen his sense of the utterance. The Folio's choice (1623) is "solid," a word that shifts the emphasis from the corrupted nature of the flesh to the impossibility of its melting into "dew." The sullied/sallied/solid choice, though small and unlikely to be noticed by the uninformed audience member, suggests a different attitude on the part of the character to his body and those of other people in ways that might speak usefully to the actor or director's sense of the character and the play. A Hamlet who is driven by a deep-seated psychological loathing of sex and physicality (on either Freudian or quasi-theological grounds) might best say "sullied." One who feels marginalized by the court and at odds with all around him— at war, as it were—might best say "sallied," if the word is deemed sufficiently intelligible. Finally, a Hamlet preoccupied with the desire for self-erasure or death might opt for "solid." These are minor but suggestive choices, and given consistent attention throughout a performance script they will have palpable effects on the production as would the bolder choice of taking readings from the First Quarto (1602).[1] Being aware of such differences before sitting down to edit the script increases the range of options for the dramaturg and thus for the production.

Another instance from *Hamlet* concerns the protagonist's exclamation of "O, vengeance!" during the "rogue and peasant slave" speech (2.2.582), a line some editions cut because it does not appear in the Second Quarto, which is often preferred over the Folio text (which does include the line), sometimes on the grounds that it feels melodramatic and doesn't scan metrically. I like the line for precisely those reasons; it builds to a high note the previous expressly theatrical language of revenge tragedy ("Bloody, bawdy villain, remorseless, lecherous, treacherous, kindless villain") inspired by the Player's presentation of the Trojan war scene immediately before, and it ends that beat nicely, its scansion suggesting a deflatory pause before "Why what an ass am I?" In short, I think the line has playable possibilities that help the speech and I would want to at least try it, even if the copy text I was working from omitted it.

Other instances of editing that might be addressed at the level of early texts include issues of line attribution. In *The Tempest*, for example, many editions ascribe the following lines directed at Caliban to Prospero:

Abhorred slave,
Which any print of goodness wilt not take,
Being capable of all ill! I pitied thee,
Took pains to make thee speak, taught thee each hour
One thing or other. When thou didst not, savage,
Know thine own meaning, but wouldst gabble like
A thing most brutish, I endowed thy purposes
With words that made them known. But thy vile race,
Though thou didst learn, had that in't which good natures
Could not abide to be with; therefore wast thou
Deservedly confined within this rock,
Who hadst deserved more than a prison. (1.2.354–65)

The vitriolic righteousness of the lines (and in these more post-colonially sensitive days their shrillness seems especially unsettling) may well seem better suited to Prospero's mouth, but the Folio assigns them to Miranda and there is no real evidence for giving them to her father beyond the (hopeful) notion that since Miranda has no similar speeches, the attribution of this one to her is merely a printer's error. Clearly a production must wrestle with such an issue, since the question of whether or not she says these lines will have a significant impact on the nature of her character and that of the production in general, as do all debates over line attribution.[2]

OPACITY AND ARCHAISM

Once a core text or texts has been selected, and the dramaturg has a good sense of the key textual choices that are rooted in early printings, she may then proceed to the crucial business of working to make the script clear. Opacity and Archaism are the bastard children of Shakespeare on stage and, as Scrooge was told of the children Ignorance and Want cowering beneath the gown of the Ghost of Christmas Present, the dramaturg should beware them both. Opacity—any utterance that is baffling to the point of incomprehensibility—is the sullen boy who will claw your show to ragged tatters. Archaism—the language of the ancient and arcane—is his sister who will shroud the production in the dust and ashes of the tomb. As the Ghost said to Scrooge, "Beware of them both, and all of their degree, but most of all, beware this boy . . ." (Dickens 107).

Of course, not everyone will agree as to what constitutes either, and in this the dramaturg is not always the best judge since he or she is already rooted in the language and thought of the Renaissance. Many times I have assumed words to be quite clear only to find that actors or audience members had no idea of what they meant. Conversely, I have, as a dramaturg, overcompensated, cutting or modernizing unnecessarily in order to bridge a gap only I thought was significant. In this, as in other things, the best route is to ask questions and float possibilities until a solution or compromise presents itself. These questions might best be asked of people who might be in the audience, rather than people involved in the show, since that is where the show's meaning is created.

Since the audience is the ultimate litmus test as to whether something "works," I am sometimes wary of actors' claims that they can take a baffling phrase and "make it work." What this usually means is that since they understand the phrase, they can become comfortable with it and will be able to deliver it naturally; but while an audience's understanding demands that an actor knows what he or she is saying, this factor alone cannot overcome truly opaque expression. The actor might be able to carry the moment so that the audience isn't scratching its head, but that is rarely enough to render the lines genuinely meaningful.

Similarly, I mistrust the impulse to put the aesthetics of sound over the content of sense. Lines that are beautiful but bewildering can have a significantly detrimental effect on the play, encouraging the audience not to listen for meaning, but lulling them into that which always haunts Shakespeare on stage, that general wash of an untethered prettiness that turns the script into the pleasing drone of background noise rather than the specifics of story and character, issue and idea. Wherever possible, of course, sound and sense should be balanced, the meaning and aesthetics (which are, after all, symbiotically bound together) working together in harmony. If one has to be sacrificed for the sake of the other, however, I would put meaning over the gracefulness of the utterance in ninety out of a hundred cases.

While it cannot be expected that every audience member will grasp the content and nuance of every line—and attempting to make it do so would drastically reduce the script's more complex utterances—it is certainly true that the cumulative effect of numerous confusing or impenetrable moments on stage is a significantly, even irreparably, damaged production.

The dramaturg's job is to see that this does not happen and he does so with an attention to detail. Will two or three baffling archaisms derail a show? No. What about twenty? A hundred? While there may be reasons to deliberately befuddle one's audience, theatre should strive most of the time for clarity, and this is achieved only with painstaking attention to anything that will jar the audience's comprehension, or lull them into the deadening sleep of a script that is all sound and fury—though possibly appealing sound and fury— signifying precious little.

The first question the dramaturg should ask when presented with a passage of text that is opaque is "why is it like this?" Is it simply a matter of our language having evolved to such an extent that this one moment is utterly mired in a lost diction, or could something else be at work? Consider, for example, this moment from Ben Jonson's *The Alchemist* in which the two chief conmen perform their virtuosity to a doubtful client.

Subtle: Sirrah, my varlet, stand you forth and speak to him,
Like a philosopher: answer i'the language.
Name the vexations and martyrizations
Of metals in the work.
Face: Sir, putrefaction,
Solution, ablution, sublimation,
Cohobation, calcinations, creation, and
Fixation.
Subtle: This is heathen Greek to you, now?
And when comes vivification?
Face: After mortification.
Subtle: What's cohobation?
Face: 'Tis the pouring on
Your Aqua Regis, and then drawing him off,
To the trine circle of the seven spheres.
Subtle: What's the proper passion of metals?
Face: Malleation.
Subtle: What's your ultimum supplicium auri?
Face: Antimonium.
Subtle: This's heathen Greek to you? (2.5.18–31)

And so on. When I worked with student actors in a production of this play they were initially baffled by the language and by my decision to keep it unedited. A modern audience, they said, would have no idea what was being said. Since I was set on keeping it, the actors considered studying the terminology of Renaissance alchemy so that they could deliver the lines convincingly. In fact, however, the lines do not depend on comprehensibility at all, quite the contrary, and while elements of the Renaissance audience may have been conversant with some of the terms in this exchange, it is unlikely that such diction would have meant much to most people. The dynamic of the scene depends not on explanation of the language then, but on the use of obscure linguistic terms as a show of learning that no one on stage (or off) is supposed to have. The doubtful religious client who believes that all languages except Hebrew are "heathen" (2.5.17) is beaten into submission by a display of rhetorical pyrotechnics. In the case of my production, the actors realized that the meaning of the words didn't matter (and despite their typically Jonsonian accuracy in alchemical terms, they may as well be gibberish), and that maximum comic effect could be obtained by rapid declamation instead of explanation. The result was hilarious, not in spite of the baffling diction but because of it.

In short, the baffling text can, paradoxically, produce a kind of clarity in the theatrical moment. In this performance of *The Alchemist*, the lines were not merely funny, they taught the audience something about the characters' ability to improvise, to bludgeon linguistically rather than explain (another Jonsonian characteristic), to overwhelm resistance by the performance of a deep-seated professionalism based in mystical/scientific blather.

In Shakespeare, whose comedy is generally less satirical, such moments are fewer than in Jonson, but there are plenty of instances in which characters are revealed by their linguistic clumsiness. Some characters speak in circular or overly verbose phrases (Polonius in *Hamlet*), while others commit grotesque malapropisms, such as Dogberry in *Much Ado About Nothing*, or misuse the language of a given professional or cultural field (law, for example, as used by *Hamlet*'s gravediggers). A similar though less comic moment occurs, perhaps, in the densely historical account of the English claim to the French throne, which occurs in the second scene of *Henry V* (1.2.33–95). The Archbishop of Canterbury's fifty-line "explanation" can be read as emphasizing not so much the legality of the ensuing war as the thin legalism of its premise. His conclusion that Henry's right to the French throne is "as clear as is the summer's sun" (1.2.85) is at least potentially ironic and playable as such. In other words, there are moments when a play's opacity is *playably* opaque. This opacity seems deliberate—and the deliberation is crucially about the production and not about the author or the text (whether Canterbury's "explanation" was intended by the author to be ironic is a moot point)— though the audience has to be let in on the joke/point, and that may still require some editing.

But what to do in the face of opacity that seems not to be constructive of itself, that does not seem deliberately intended to make a point, or whose joke still needs some clarification to be playable? My instinct in the first pass on the script is to begin with the more conservative options before progressing to more adaptive strategies, and in the face of difficult or obscure phrasing, I would recommend that the dramaturg begin by looking at early textual variants. These will not always solve the problem, of course, especially since the work done by modern editors has generally already selected the "best" or simplest choice, but they sometimes can raise playable options, particularly when the text the dramaturg is working from is old or flawed, or the play is derived from competing early printings that are substantially different (as in the case of the folio and quarto texts of *King Lear*).

Where the earliest printings are unhelpful, the next step is to examine the difficult lines for a thorough sense of their meaning as text, so that their worth for the production in terms of both sentence-level meaning and thematic or poetic nuance can be assessed. This evaluation of the opaque passage might be rendered as the following series of questions.

1. Do the lines reveal important elements of the plot? If they contain plot information but that information is stated elsewhere, do these lines clarify the plot point or remind the audience of something they may have otherwise forgotten?
2. Do the lines develop valuable issues of character?
3. Do the lines contribute valuable mood? What is the "genre" of the moment, or are there facets of multiple genres? For example, do lines that are ostensibly comic contain darker elements that make a more serious point or connect to larger issues elsewhere in the play? If they are supposed to be funny, are they? Is the joke worth saving?
4. Do the lines present or re-present significant ideas, themes, or images that collectively define an important aspect of the play?
5. Are the lines particularly pithy or resonant?
6. Are the lines famous? Answering this does not offer a clear indication of what the dramaturg should do, but she should at least be aware that cutting or tampering with lines that are well known, however baffling they might actually be, may well produce disappointment or irritation in the company or audience.
7. In general, what is our audience—insofar as we can gauge such a complex and disparate thing in advance—likely to get out of the lines as written? If the lines are worth saving, does our sense of the audience suggest any strategies we might use to effect that saving?

Having answered these questions, the dramaturg is better equipped to approach the matter of cutting, modifying, or transforming the words utterly, and the various degrees and combinations of these strategies. The decision in each case will be affected by the general adaptive philosophy of the production that was worked out with the director, but will largely be determined— as the third question suggests—by the energy and "genre" of the moment as it might work in performance.

Comedy, for example, feels more disposable than tragedy, because it is assumed to be less important (a fact that is evident in the way critics assess film and television as well as Shakespearean drama), and there is no doubt that lines that are supposed to be simply comic but are not funny anymore need to be modified or cut. Few things are more annoying than actors delivering impenetrable lines and laughing uproariously to signify that they are saying something funny. Stage business can, of course, clarify (or carry) some jokes, but there's only so much pelvic thrusting one can stand on stage. The problem, as questions 3 and 4 suggest, of course, is that Shakespearean humor is rarely just humor, and in the last 400 years critics have managed to find ways of seeing serious purpose in the most apparently throw-away of quips. As ever, the dramaturg must rely on his judgment, using the above questions to ascertain the value and importance of the lines before determining a strategy for cutting them or finding other methods of making them work. In matters of humor, the dramaturg has the doubly difficult task of being both clear and funny, particularly where the lines are not loaded with more obviously "serious" purpose. Consider this brief exchange from *Henry IV Part 1*.

> **Prince**: Thou judgest false already. I mean, thou shalt have the hanging of thieves, and so become a rare hangman.
> **Falstaff**: Well, Hal, well; and in some sort it jumps with my humor as well as waiting in court, I can tell you.
> **Prince**: For obtaining of suits?
> **Falstaff**: Yea, for obtaining of suits, whereof the hangman hath no lean wardrobe. 'Sblood, I am as melancholy as a gib cat or a lugged bear.
> **Prince**: Or an old lion, or a lover's lute.
> **Falstaff**: Yea, or the drone of a Lincolnshire bagpipe.
> **Prince**: What sayest thou to a hare, or the melancholy of Moorditch?
> **Falstaff**: Thou hast the most unsavory similes, and art indeed the most comparative, rascalliest, sweet young prince. (1.2.64–80)

Funny stuff. The problem with playing a role like Falstaff, of course, is that while he is famously funny, his brand of humor, like a lot of Shakespeare, is verbal, tied to puns and a certain delighted playfulness rooted in language. Not only is this kind of humor less common these days and so can feel stale and labored, English itself has evolved, the cultural context has been lost (*the melancholy of Moorditch?*) and much of what thus made Falstaff funny (and, worse, famously so) is thus incredibly hard to pull off on stage. The audience expects to be amused but often feels merely confused, however much everyone on stage is falling about and slapping their thighs at the hilarity. Clearly, something has to be done, though whether that is cutting or modifying (or a combination of the two) will depend on the show. What the dramaturg can't do is leave the actors hanging out there, saying this stuff as the crickets chirp and the tumbleweed rolls by. This is also true of comic moments in tragedies. The dramaturg can't hide behind the play's overall weightiness when the funny moments aren't: comedy is in the genre of the moment, not the production as a whole.

Tragic moments tend to be permitted more license in terms of opacity (particularly if they are well known), because they ooze a general aura of seriousness even if they are not clearly intelligible. While it is certainly true that the dramaturg has to be alert to what the audience expects of the show, she should resist the sense that any impenetrable utterance should be allowed to stand because it seems weighty or is familiar. Consider, for example, these lines from some of the most famous pieces of Shakespeare, soliloquies by Hamlet and Macbeth. In each case, I have italicized words or phrases that I suspect many audience members would be hard-pressed to translate into their own idiom or whose sense might not be completely clear. First, from Hamlet's "To be or not to be" speech:

There's the *respect*
That makes *calamity of so long life.*
For who would bear the whips and scorns of time,

Th' oppressor's wrong, the proud man's *contumely*,
The pangs of disprized love, the law's delay,
The insolence of office, *and the spurns*
That patient merit of th' unworthy takes,
When he himself might his *quietus* make
With a bare *bodkin*? Who would *fardels* bear . . . ? (3.1.69–77)

And this convoluted sentence from Macbeth's speech before killing Duncan:

If th'assassination
Could *trammel up the consequence*, and catch
With his *surcease success*—that but this blow
Might be the be-all and the end-all!—*here*,
But here upon this bank and shoal of time,
We'd jump the life to come. (1.7.2–7)

There are, of course, good reasons for the difficulty of the speeches, and it could be that the dramaturg will want to leave them intact for equally good reasons having to do, for example, with the tortured state of the speakers' minds. What I would insist upon, however, is that such lines are not left alone solely out of a sense of reverence, either for their fame, or for their generalized aura of seriousness, be that philosophical or psychological. Sometimes clarifying or editing around these kinds of lines can make the audience see their pith more clearly, and may even convey *why* the lines are worthy of their fame.

Some dramaturgs assume that awkward or opaque lines have to be either cut or left as they are, an assumption premised on the idea that introducing new words (the dramaturg's) is more adaptive than cutting the original. This is only partly true, excision being at least as adaptive as alteration, and I'm wary of that binary ("keep it as it is or cut it"), which seems simplistic and reductive. Some lines can be saved with the addition of a new word, something that is surely more desirable even to the textual purist than axing the lines outright.

One editing rationale thus presents itself: the dramaturg can save elements of the original by updating or otherwise editing the text, maintaining a "spirit" of the original utterance at the expense of literal adherence. Take the lines from *Hamlet* I cited above, for example. Some of the words there—*contumely, quietus, bodkin*, and *fardels*—will baffle many audience members, and though that audience may be able to extrapolate the sense of the words from their context, they could be replaced with others that approximate their sense but are clearer in performance. Any word substitution has to be made cautiously, of course, and, as much as possible, with an eye on the knowledge and assumptions of the specific audience who will see the show. Issues of clarity have, moreover, to be balanced with issues connecting to the integral worth (be it semantic—what the words actually meant when written, or cultural—what the audience expects to hear) of the original wording.

As I said in chapter 9, we stage Shakespeare today for many reasons, but one of them is faith in the value and power of the words themselves. Altering these words to make the current production work as theatre must therefore weigh the issue of clarity against the richness of the textual original, and while a production will quickly adopt a generalized sense of its adaptive range, word substitution must be made on a case-by-case basis. While it is relatively easy to take a position at either end of the adaptive spectrum ("we change whatever we want on a whim" or "we insist on using the First Folio unedited"), the results tend to miss opportunities for the production of a richer theatrical product. I do not believe that there are short cuts to good editing, and while all decisions are ultimately subjective (both in terms of *whether* a change should be made and *what* that change should be), all choices should thus be made with a clear sense of the original's worth and complexity. It may be that clarity for all has to be occasionally shelved in favor of richness for some, so long as that *some* doesn't really mean *the dramaturg*.

That said, the difficulty in making word substitutions in the Shakespearean text is often about finding suitable words, and the acceptability of the practice is often thus dependent on its execution. All alterations are not created equal, the good choice suiting in terms of both sound and sense, even if they do not mirror exactly the word that has been removed. In the lines from Hamlet's soliloquy cited above, some substitutions are easier than others: for *contumely* try "arrogance" ("insolence" is better, but it appears two lines later), for *fardels*, try "burdens." *Quietus* is tricky, and though I have heard it replaced with "hiatus" I would be tempted to try something like "acquitance" (after David Bevington's footnote: footnotes in contemporary editions can sometimes be a good source for word substitutions). *Bodkin* worries me less, because the word can easily be accompanied by the appearance of a physical dagger, so its meaning is quite clear, and I can think of nothing that will fit the meter of the line and maintain the trivializing alliteration with "bare."

The weapon reference, however, points up a common instance in which productions seem to demand alterations to the script, which is when characters refer to swords but brandish knives and firearms, often to the momentary amusement of the audience.[3] As I said in chapter 2, the theatre's allegiance is to its own artistic and creative power, not to a slavish replication of the text, and when replication of that text is brought into nonconstructive dissonance with the theatrical moment, some kind of change should be made to the script. The first rationale for adding words to the script (reshaping the "spirit" of the textual original, which might otherwise be lost) thus gives way to the second: the principle that theatre authorizes itself. Each production is a new work of art, each one is unique, and to shackle the production's ability to occasionally modify the script, even in a relatively nonadaptive, textually conservative production, is to stifle those impulses and reduce the theatrical moment to a mere parading of the text. Finally, it puts the mistiness of high cultural nostalgia above communicative clarity in ways seriously calling into question the purpose of staging the play at all.

Cutting or refashioning is particularly called for in the face of impenetrable contemporary reference and allusion. Consider, for example, the following example from *Henry V* in which the Chorus conjures the exaltation of the

English people on the king's victorious return from France:

> But now behold,
> In the quick forge and working house of thought,
> How London doth pour out her citizens!
> The Mayor and all his brethren, in best sort,
> Like to the senators of th' antique Rome
> With the plebeians swarming at their heels,
> Go forth and fetch their conquering Caesar in;
> As by a lower but loving likelihood,
> Were now the General of our gracious Empress,
> As in good time he may, from Ireland coming,
> Bringing rebellion broached on his sword,
> How many would the peaceful city quit
> To welcome him! (5.0.22–33)

The image from Roman history gives way to an immediate, contemporary reference generally taken as invoking the earl of Essex ("the General") and his (unsuccessful) campaign in Ireland against Tyrone in 1599 on behalf of Queen Elizabeth ("our gracious Empress"). Only the best-informed audience members will understand this latter reference, and even they will not hear it as the original audience might have, for whom that campaign was a matter of pressing concern and—possibly—controversy. Productions have often omitted these lines, though they are sometimes emended to reflect a contemporary event, though some audiences balk at such a change, as did one reviewer of a 2003 New York production:

> Occasionally things in the production don't work, especially some of the topical references, for instance a Princess who imitates Marilyn Monroe's breathy "Happy Birthday, Mr. President" and the inserted line for Chorus, "our President from Baghdad coming with Saddam skewered on his sword."[4]

What "works," of course, is a highly subjective matter, and some reviewers and audiences will dismiss any such deliberately topical reference as a bastardizing of the text, though the spirit of such emendation is a good deal closer to Shakespeare's original than a reference to an obscure historical event that has long since ceased to resonate for the audience. The dramaturg should have a clear sense of the production's tonal agenda before making such a decision, and should do so in concord with the director and—in the case of adaptive strategies that may draw fire from the audience or critics—the company manager or artistic director. Again, this is not a matter of adaptation per se (since all production is adaptation) or even of tinkering with the text (something I advocate should be done frequently). It is a question as to how markedly the production wants to *point out* the degree of its adaptive strategies. Cut the problematic lines about Essex and few audience members will even notice the change; add a reference to Saddam Hussein and you put the adaptation issue center stage, complicating it further by the visceral and divergent responses that the subject of U.S. foreign policy is likely to produce within the audience. In general, of course, cutting is safer and avoids the accusation of being heavy-handed in its indications of contemporaneity, but more radical adaptation in such instances validly seeks to jar the audience out of its composure, both in terms of its attitude to the immediacy of the play on stage NOW (not in history) and to problematic assumptions about textual "purity."

The *Henry V* example is a relatively rare moment in Shakespeare in which a character (if we can call the Chorus a character) exhorts an audience directly to think in terms of its own immediate cultural moment, but the number of times contemporary issues or events are touched on less expressly—but still in ways the original audience would have found strikingly imminent—are too numerous to mention. The Porter scene in *Macbeth*, for example, in which the porter imagines he is the gatekeeper of Hell, is crammed with references the Renaissance audience would have connected with immediately, but that are largely lost on a modern audience:

> Knock, knock, knock! Who's there I' the name of Beelzebub? Here's a farmer that hanged himself on th' expectation of plenty. Come in time! Have napkins enough about you; here you'll sweat for't. Knock, knock! Who's there, in th' other devil's name? Faith, here's an equivocator, that could swear in both the scales against either scale, who committed treason enough for God's sake, yet could not equivocate to heaven. O, come in equivocator. Knock, knock, knock! Who's there? Faith here's an English tailor come hither for stealing out of a French hose. Come in, tailor. Here you may roast your goose. (2.3.3–15)

Footnotes in any good edition will unpack the content of these arcane references, but the stage is no place for footnotes or for what GSF has come to refer to as the Pop-Up Dramaturg.[5] Unless the production's audience is very unusual (and probably academic), the communicative value of these lines is fairly minimal and actively works against the largely (albeit darkly) comic impulse of the scene. In this case, I would advocate taking a strong hand, either cutting most of the lines outright or, if the ambience of the show permitted it, altering the lines so that they would be clearer and, probably, more contemporary. There are, after all, plenty of contemporary examples of unscrupulous professionals who might stand in for the farmer and tailor. I would, however, do what I could to retain the reference to the equivocator, not out of concern for the purity of the text or for history (these lines appear to refer to the trial of Jesuit priest Henry Garnet for his involvement in the gunpowder plot, and help to date the play as originating in 1606), but because equivocation more generally is a major theme throughout the play, manifested in particular by the ambiguous utterances of the weird sisters, the testing of Macduff by Malcom, and by Macbeth's own syntactically and morally tortured soliloquies.[6]

While instances of direct contemporary reference are numerous in the plays, they are less frequent than more indirect references to belief systems, bodies of thought, assumptions about society and culture, and other matters that saturate the period and thus the plays in ways that are so total that the idea of stripping them away from the text effectively calls for the complete rewriting of the script. While some productions will—in the spirit of extensive adaptation—take that very approach, most want to retain the plays as written as much as possible, thereby producing a logical and practical problem.

Let us stay with *Macbeth* for a moment. The play on stage is often discussed simply in modern psychological terms, but the text is obviously shot through with ideas that are quite foreign to a modern audience and which significantly alter that sense of psychology. The murder of Duncan, for example, is not simply a murder, but a regicide, a crime with very different associations for a Renaissance audience, a crime inflected by its place in various political and theological models of the day. Similarly, witchcraft is, for many people in the Renaissance, a real and potent presence, as are other burning issues in the play: a Calvinistic sense of predestination, a clear and limited sense of the nature and role of women, and a complex and contradictory set of ideas about tyranny and how people are supposed to counter it. These issues, issues that dominate the play, define it even, are largely alien to the modern audience experiencing the play on stage, and they therefore pose significant problems for the dramaturg who is trying to make the play "work" while still retaining aspects of the text's core (and thus historic) nature. The key, I think, is to remember that the production is neither the text nor the play and it cannot be expected to re-create the original audience's experience. The dramaturg can, however, be alert for these large-scale ideas and issues, and give them special attention when they surface in the script, so that they might be addressed in ways that balance clarity and depth, through a judicial use of cutting and changing, perhaps through connecting them to more contemporary issues by analogy (and here, as elsewhere, I use the word "analogy" to mean that which is similar, not that which is identical). It is also, as I have said, a crucial aspect of the dramaturg's job to recognize when certain foreign or arcane ideas are simply not playable on the contemporary stage, and that extensive script work to make them so is likely to be time wasted.

There are times, of course, when the altering of archaic words or phrases is not worth the effort and can be counterproductive. For example, I generally leave antiquated pronouns like "thou" or "thee" because I assume that even if the members of the audience don't know what such words mean (and almost all of them will), the context of the delivery will generally make them work. Likewise, I leave archaic verb forms ("doth," "sayeth," and so on) because their sense is usually quite clear and, while they may be unfamiliar to the audience, they are not actually opaque. In the initial script preparation, then, I would leave them, always assuming that further changes could be made during rehearsal if certain words or passages prove difficult either to say or to understand. Though I'm wary of placing any barrier between the audience and total comprehension, I don't think mere unfamiliarity constitutes such a barrier, and modernizing words like this leaves the production open to charges of inauthenticity; audiences expect a certain historical distance from Shakespeare's language, and however much dramaturgical polishing might render that language clearer, it can also disrupt the audience's (sometimes problematic) sense of getting the genuine article in a fairly unadulterated form. Clarity does not, after all, demand immediacy, and it is important to remember that part of what makes the plays worth doing is their language. While such an assumption does not render the text untouchable, it does suggest that it is easy to throw out the baby with the bathwater if one edits with a view to rendering the plays *absolutely* clear or contemporary in sound and sense.

The more laissez faire attitude such a position suggests gets more tricky when approaching lines that are familiar, even famous, and this is an important counterpoint to my earlier insistence on clarity in such cases. Most people, for example, do not know exactly when the Ides of March occur, but if a soothsayer in a production of *Julius Caesar* tells Caesar to beware March 15, the show is likely to be ridiculed. Maintaining the original "Ides" over something clearer can be justified in various ways (not least of which is the sheer sound of the open vowel at the heart of "Ides"), but the best reason is that the line is well known, is part of history and culture, one of those quotable moments whose resonances go beyond the stage and beyond mere semantics. Some might say that the same is true of the words I suggested changing in Hamlet's soliloquy, but I would argue that the cases are different. When exactly the Ides of March are doesn't matter. The word suggests a specific time and the play announces when that time has arrived; *Ides* might be unfamiliar, then, but context indicates its function if not its precise sense. Words like *contumely* and *fardels*, by contrast, depend on their semantic value and, since the speech is less memorized in school than it used to be, they don't carry the same broader resonance that *Ides* does.

Part of the decision about whether or not to substitute words is therefore about rendering a production clear in ways that do not reduce it significantly and that do not violate audience expectation in ways that will invite charges of "dumbing down" the play or of pandering to the lowest common denominator. Translation into English of the phrase "et tu Brute" would surely render the utterance clearer, but would also show the production's adaptive hand in ways bound to unsettle some audience members who may be able to quote that line alone from the play.[7] The problem with changing famous lines—even if they are rendered clearer or otherwise better—is that the change unavoidably announces itself to the audience. While there may be good reason for doing this, the production risks creating precisely the kind of distraction that the substitution was designed to avoid in the first place. Therefore, while it is essential that the dramaturg scrutinize every line for words and phrases that may cause bafflement or incomprehension, the impulse to change such phrases must always be balanced against a sense of how easily, convincingly, and invisibly—or rather inaudibly—the change can be made.

Thus far I have framed the discussion of editing the script in terms of lines or ideas that feel opaque or arcane, but there are circumstances in that the text contains words that seem quite clear but whose meaning or nuance has shifted significantly. When Lysander in *A Midsummer Night's Dream* says that Demetrius "made love" to Helena (1.1.107), for example, he probably means something like "courted" or "wooed," rather than "had sex with," which is what a modern audience might assume. While a production may see value in exploiting this slippage into a more sexualized register, it might be helpful for the dramaturg to suggest an alternative such as "spoke love," a slightly cumbersome phrase, perhaps, but one that most audience members will think feels vaguely Shakespearean (and metrical) and will thus pass without comment. In rehearsals for a production of *Cymbeline*, the actors laughed every time Cloten dismissed the rural inhabitants of Wales as "mountaineers," because the word evoked a modern sense of recreational climbers, even (worse) skis and liederhosen! We struggled without success (and there were audience chuckles at the word throughout the run of the show) to find an alternative phrase that would fit the meter of the line, one that would convey the core elements of the term in Cloten's mouth, particularly a sense of rustic debasement (which is much more important than the link to the mountains themselves). As such we tried simply "villagers" and "mountain men," though other less literal phrases might have worked better, "rustic fools" or "country clowns," perhaps, words out of Shakespeare's own idiom, which would be clear in sense but would not stand out from the surrounding script when spoken aloud.

Other instances requiring some kind of translation might be more literal, as when characters in Renaissance plays pepper their speech with phrases (usually proverbial) from other languages. In general, the dramaturg is best to translate such phrases or, if their content is contained elsewhere in the speech, as is often the case, cut them outright. There are exceptions to this principle, of course, as when the phrase is still familiar (Caesar's "et to Brute," for example), when the wording is usefully obfuscatory, or when the content of the phrase is less important than its speaker's erudition or pompousness. When working on a production of Kyd's *The Spanish Tragedy*, I retranslated the final play within the play into the foreign languages in which, according to a note in the original printing, it had first been staged, in order to replicate the baffling sense of linguistic chaos, an idea the production had been at pains to play up throughout the show. Another exception to the translation rule is when foreign language is used to point up a more general foreignness (sometimes for comic effect) as is the case in the so-called English lesson scene in *Henry V*, in which the French princess' struggles to learn English lead her into various smutty jokes.

Several times I have suggested that there is value in maintaining the metrical pattern of the lines when one cuts or changes the text. There are five reasons for this. One is aesthetic, the sense that the meter creates a sense of balance and rhythm, which adds to the beauty and thrust of the line. The second is that the (generally) iambic pentametrical line serves as a mnemonic for actors, helping them to remember their lines and keeping them on track if they stumble on a word or two. Third, sound and sense are symbiotically bound, and to change the feel of the line thus alters its meaning. While there may be good reason for doing this, the dramaturg should at least be aware that a change to the aural pattern of the line is a change to sense as well.

The fourth reason to be wary of disrupting the verse's metrical arrangement is made much of by actors and directors committed to an approach to Shakespeare (particularly in the United Kingdom) that is rooted in verse speaking: that the rhythm of the line contains actorly information about emphasis, intellectual and emotional focus and breathing. Much of this is undoubtedly useful, though I am suspicious of the way such valorizing of the script suggests intimations both of correctness and textuality. Even if we can assume that the printed text (invariably the Folio) does contain such clues for delivery, it doesn't follow that how the lines *were* done (and we cannot reconstruct that delivery precisely) dictates how they should be done now. "Again, today's audiences are not Shakespeare's and it is to today that the production is addressed, a fact which should always be borne in mind however much we value the use of the lines and their meter as a guide for performance."[8]

The centrality of the audience to the construction of meaning, however, presents a fifth reason for why the dramaturg should take pains to maintain the meter. The audience hears the meter of the verse if it is passably delivered and, more to the point, they hear when it breaks. Often Renaissance playwrights will use interruptions in the meter of a line to indicate a shift either in pace or in concern, a moment that modern actors can often use psychologically to shift gears. But when a break in the meter occurs because of a cut or change in the text made by the company, that cut or change is announced to the audience by the shift in the meter. This can create a momentary sense of disconnect and can make the change to the script jarring, even—again— raising accusations of inauthenticity that might otherwise have been sidestepped. As the dramaturg should pick her battles with the director and actors, so she should with the audience, and minimizing the impact of changes by maintaining the verse form is a way of avoiding charges of clumsiness and interpolation.

When altering a line, then, the dramaturg should examine the metrical pattern of the entire passage, not just the problematic phrases, in order to *hear* what the rhythm is doing around the words that are to be altered. Where the entire passage is regular, the altered version should also be regular, where the meter of the larger passage is inconsistent or erratic, the regularity of the altered phrase is less pressing, even if the replaced words were themselves metrical. While these principles mean that prose is always easier to edit than verse, they also mean that verse which is edited well is less likely to be spotted by the audience, since the meter tends to convey a sense of organic participation. In short, maintaining the meter of the original is more likely to make the new words seem to "belong." If this seems disingenuous the dramaturg should remember that what is at stake here is in part a matter of being held up as inauthentic, a charge that, as part I of this book has already said, is largely spurious, however widespread such assumptions may be.

A similar strategy open to the dramaturg is the reassignation of lines. As with other sentence-level alteration this needs to be handled cautiously, but it can be a useful way to clean up issues that stem from modern actorly concerns and notions of character, as well as possible errors or ambiguities in original texts. For example, consider the following example from *A Midsummer Night's Dream* in which the mechanicals plan their production.

SNOUT
Doth the moon shine that night we play our play?
BOTTOM
A calendar, a calendar! look in the almanac; find out moonshine, find out moonshine.
QUINCE
Yes, it doth shine that night.
BOTTOM
Why, then may you leave a casement of the great chamber window, where we play, open, and the moon may shine in at the casement.
QUINCE
Ay; or else one must come in with a bush of thorns and a lanthorn, and say he comes to disfigure, or to present, the person of Moonshine. Then, there is another thing: we must have a wall in the great chamber; for Pyramus and Thisby says the story, did talk through the chink of a wall.

While the passage is perfectly clear (the only word requiring change, I think, is "lanthorn," which is easily amended to "lantern"), a production may want to throw greater emphasis onto the character of Bottom, whose role in the play is much more significant than that of his fellows, including Quince. By reassigning some of the lines, a clearer and more consistent dynamic can be achieved, which, though clearly a rewriting of the original, produces a result that might be more in keeping with the characters as the production wants to play them, particularly emphasizing Bottom's dominance over the group, thus:

SNOUT
Doth the moon shine that night we play our play?
BOTTOM
A calendar, a calendar! look in the almanac; find out moonshine, find out moonshine.
SNOUT
Yes, it doth shine that night.
QUINCE
Why, then may **we** leave a casement of the great chamber window, where we play, open, and the moon may shine in at the casement.
BOTTOM
Ay; or else one must come in with a bush of thorns and a lantern, and say he comes to disfigure, or to present, the person of Moonshine.

QUINCE
>   Then, there is another thing: we must have a wall in the great
>   chamber; for Pyramus and Thisby says the story, did
>   talk through the chink of a wall.

The result puts Bottom both in charge of and a little outside the rest of the group (I also changed "you" to "we" in Quince's first speech here), a position emphasizing his desire to run things in ways that are both absurd (he inherits Quince's malapropism "disfigure") and impressive, at least to his fellows. While this is a fairly minor alteration (and one that goes largely unnoticed by even the informed audience member), it gives the actors a slightly different sense of how they might see themselves and how they relate to each other. Such minor changes can clarify an actor's sense of character in ways that make for a stronger, more consistent performance, or may be tailored to a character dynamic that emerges from the actors themselves and is, admittedly, non-textual in origin.

Another way of altering the text in ways that are close to invisible in performance is through the manipulation of that which appears on the page but is unsaid, namely, stage directions and punctuation, both of which tend to be of questionable authorial status and are treated by most actors as flexible. The dramaturg can make use of these facts in the production of the script by scrutinizing issues such as the timing of entrances and exits; is Hamlet's "To be or not to be" speech a soliloquy or a monologue delivered to Ophelia? At what point, if at all, does he suspect he is being spied on by Claudius and Polonius in that same scene?

Clearly the writing or rewriting of stage directions into the script may be seen as invading the territory of the director, or as overdetermining the rehearsal process, but such things presented more tentatively either in the script or as notes to the director can open up fruitful possibilities. One such is in the matter of asides, which are rarely clarified as such in the earliest printings of the plays, but that have significant implications for issues of individual character and group dynamics. Such things are flexible, and while the dramaturg's script does not decide such matters (they are subject to directorial review and, more importantly, the exploratory conditions of the rehearsal room) it can get possibilities that would otherwise be passed over into circulation.

In matters of punctuation the dramaturg has more range, since the actors will expect punctuation to be clearly marked in the script at least provisionally, and in this the dramaturg's choices can get some options into play early. Again, punctuation in Renaissance texts tends to be erratic, but can appear prescriptive on the page in unhelpful ways that affect the delivery of the lines. As ever, the dramaturg's first recourse is to punctuate in ways that make sense, correcting flawed texts and availing himself of modern textual editions that will outline a range of possibilities. The dramaturg may also choose (either during initial script preparation or during rehearsal) to raise new possibilities. One that is often productive is to ask whether a line given as a statement might be phrased as a question, even a rhetorical question, or the other way round. For example, when Lady Macbeth is persuading her husband to proceed with the murder of Duncan she responds to his question, "If we should fail?" with "We fail? / But screw your courage to the sticking place / And we'll not fail" (1.7.59–61). The Oxford edition replaces the Folio's question mark after Lady Macbeth's "We fail" with an exclamation point, producing an altogether different sense of the exchange, one that accepts failure as a possible result but is prepared to risk it anyway. The question mark, on the other hand, suggests incredulity and potentially throws the emphasis onto the pronoun: *we* (of all people) fail?

As I have said, the performance script should be constructed in multiple passes so that the dramaturg and director have time to review each new set of changes (clearly marked on hard copy, or put in bold script in electronic texts so that all revisions stand out) as they are made. I generally assume that three complete passes will be made, first dealing almost exclusively with issues of sentence-level clarity and any large-scale directorial mandates that are fundamental to the production. This may include contingencies dependent on casting decisions (including doubling and role conflation) if such decisions have already been made. If the script gets assembled—however tentatively—before casting takes place, it may assist in clarifying that process.

In the second pass (after the first has been approved, presumably with alterations, by the director), I concentrate on those cuts and changes that, though not essential for clarity, make the script better suited to the production that is starting to take shape on the drawing board. This pass makes greater use of the preliminary conversations with the director and designers about what this show is likely to be about, what anchors it, what are its major areas of exploration, its intellectual and emotional currents, cutting and editing to foreground the production's dominant concerns. In the case of long plays (or of productions whose elements are likely to make the show feel long), this is the stage at which cuts driven by issues of running time might be addressed.

Running time will vary, of course, according to the nature of the production, but some rules of thumb can be helpful in initial estimation of how much of a script needs trimming to fit desired performance time. One way is to estimate running time by line count, assuming approximately six seconds per line. A glance at the line total of any printed edition is a useful start (the Brubaker method estimates six seconds of playing time per line), but another method recently proposed by Scott Kaiser uses an average word count of 130 words per minute thus providing a gauge of how long an *edited* script is likely to run.[9] This is particularly useful if one is working from an electronic text with a program such as Microsoft Word, which will perform an accurate word count automatically. Such methods are, of course, extremely tentative since all shows run at different paces, but they can be of use in estimating running time before the production is ready to start putting large chunks of the show on its feet.

The third pass (again, after the second has been approved) is a polishing stage where the dramaturg should reread the script in its entirety looking for inconsistencies either originating in errors in the original or in changes that have already been made. Audiences are unlikely to be unduly thrown by shifts in the ages of characters or whether a character is a count or a duke (one such inconsistency concerns Orsino in *Twelfth Night*), but actors like to be sure of these things as they begin their process, and some inconsistencies can have more far-reaching implications, particularly if a cut or alteration has created a logical problem or plot confusion elsewhere in the play. Likewise, the dramaturg should keep a close eye on who is on stage, when they go off, and other issues that could cause problems if they conflict with the script as written. Some of these may not become apparent until rehearsals have begun, but the smoother and more coherent the script is beforehand the better for both the actors' ease of use and the dramaturg's credibility.

This third pass should also pay particular attention to previous changes in the script to ensure that they incorporate the best choices possible, something to be assessed according to matters of aesthetics and tone as well as the clarity of the semantic content. This "final" polishing is a particularly good time to test the script in terms of meter. When this version has been approved by the director, it can go to the printer, and though this does not end the editing process, it should mark the point at which the bulk of the script work has been completed. If the job has been done well, the dramaturg has already made an invaluable contribution—perhaps her most valuable contribution—to the final production.

13

SCRIPT-EDITING EXAMPLES

In this chapter, I present some specific instances of editing for performance, attempting to provide both concrete examples and the rationale behind them. They are, of course, in no way definitive, and readers may disagree with my choices. Such disagreement is necessary since each production is innately different, and my choices have been made with the conditions of specific GSF productions in mind. The only disagreement with which I would take issue is that which claims my choices to be in violation of the original text for reasons I have already outlined. Indeed, I consider what follows to be derived from a fairly conservative notion of adaptation.

It is risky to present actual examples like this, since much that can be agreed upon in the abstract becomes divisive when given concrete form, but theatre is very much about such concreteness—actors, after all, have to actually say the words—so at some point academic or theoretical generality must give way to material specificity. On paper these material specifics look bald and conspicuous, though I think that on stage they remain so to only the most studied of audiences, and such people are not, for the most part, a production's target audience. Who that audience is likely to be will be of paramount significance in determining the appropriate editorial strategy for any given production, and my sense of our audience at GSF clearly informs some of my choices here. There can, after all, be no "ideal" script because of theatre's essentially local nature." I have no doubt that there are other and better ways of editing the passages that follow, but they illustrate a method that builds on the theoretical principles described earlier.

I have worked from scripts compiled for actual productions, though I have made further changes that either did not occur to me at the time or did not get approval from the rest of the company. That they are derived from production scripts, however, means that the work they represent was necessarily collaborative. The choices therein reflect input from actors and directors who sometimes (particularly in the example from *Julius Caesar*) brought some ideas to the table for which I was more sounding board than origin. Since the dramaturg's job is, in part, to help think through the logic and impact of other people's choices as well as his own, I have included such ideas as they were or might be manifested by the script.

The form in which the examples are laid out varies, moving from a narrative model explaining each choice and weighing alternatives as I move through the selected passage, to a less expository method in which the original and edited versions are juxtaposed against each other, moving finally to one that presents only the edited product, using footnotes to detail both the original text and any explanation that is not fairly self-evident.

EXAMPLE 1: COMIC DIALOGUE: *THE TAMING OF THE SHREW*, 2.1

I have included below the entire first exchange between Kate and Petruchio (2.1, 182–277), much of which, if not semantically absolutely clear, is at least playable. I have italicized the words or phrases that I think some actors and audiences may find unfamiliar or unclear but which I think can be made to work well-enough in the delivery. I have underlined the sections that I think are more problematic and added bold face to those that I think need to be extensively changed or cut. After the original passage, I have broken the exchange into sections and presented some of the problems of turning the textual original into a working script, suggesting possible solutions in the course of the analysis. The production which first motivated this work was directed by Richard Garner at GSF in 2002.

PETRUCHIO
Good morrow, Kate; for that's your name, I hear.

KATHARINA
Well have you heard, but something hard of hearing:
They call me Katharina that do talk of me.

PETRUCHIO
You lie, in faith; for you are call'd plain Kate,
And bonny Kate and sometimes Kate the *curst*;
But Kate, the prettiest Kate in *Christendom*
Kate of Kate Hall, my super-dainty Kate,
For dainties are all Kates, and therefore, Kate,
Take this of me, Kate of my consolation;
Hearing thy mildness praised in every town,
Thy virtues spoke of, and thy beauty sounded,
Yet not so deeply as to thee belongs,
Myself am moved to woo thee for my wife.

KATHARINA
Moved! in good time: let him that moved you hither
Remove you hence: I knew you at the first
You were a moveable.

PETRUCHIO
Why, what's a moveable?

KATHARINA
A join'd-stool.

PETRUCHIO
Thou hast hit it: come, sit on me.

KATHARINA
Asses are made to bear, and so are you.

PETRUCHIO
Women are made to bear, and so are you.

KATHARINA
No such <u>jade</u> as you, if me you mean.

PETRUCHIO
Alas! good Kate, I will not burden thee;
For, knowing thee to be <u>but young and light</u>—

KATHARINA
Too light for such a *swain* as you to catch;
<u>And yet as heavy as my weight should be</u>.

PETRUCHIO
<u>**Should be! should—buzz!**</u>

KATHARINA
<u>**Well ta'en, and like a buzzard**</u>.

PETRUCHIO
<u>**O slow-wing'd turtle! shall a buzzard take thee?**</u>

KATHARINA
<u>**Ay, for a turtle, as he takes a buzzard**</u>.

PETRUCHIO
Come, come, you wasp; i' faith, you are too angry.

KATHARINA
If I be waspish, best beware my sting.

PETRUCHIO
My remedy is then, to pluck it out.

KATHARINA
Ay, if the fool could find it where it lies,

PETRUCHIO
Who knows not where a wasp does
wear his sting? In his tail.

KATHARINA
In his tongue.

PETRUCHIO
Whose tongue?

KATHARINA
Yours, if you talk of tails: and so farewell.

PETRUCHIO
What, with my tongue in your tail? nay, come again,
Good Kate; I am a gentleman.

KATHARINA
That I'll try.

She strikes him

PETRUCHIO
I swear I'll cuff you, if you strike again.

KATHARINA
<u>So may you lose your arms:</u>
<u>If you strike me, you are no gentleman;</u>
<u>And if no gentleman, why then no arms</u>.

PETRUCHIO
A herald, Kate? O, put me in thy books!

KATHARINA
<u>What is your crest? a coxcomb?</u>

PETRUCHIO
<u>A combless cock, so Kate will be my hen.</u>

KATHARINA
<u>No cock of mine; you crow too like a craven.</u>

PETRUCHIO
Nay, come, Kate, come; you must not look so sour.

KATHARINA
<u>It is my fashion, when I see a **crab**.</u>

PETRUCHIO
<u>Why, here's no **crab**; and therefore look not sour.</u>

KATHARINA
There is, there is.

PETRUCHIO
Then show it me.

KATHARINA
Had I a glass, I would.

PETRUCHIO
What, you mean my face?

KATHARINA
<u>Well aim'd of such a young one.</u>

PETRUCHIO
<u>Now, by Saint George, I am too young for you.</u>

KATHARINA
<u>Yet you are wither'd.</u>

PETRUCHIO
<u>'Tis with *cares*.</u>

KATHARINA
I care not.

PETRUCHIO
Nay, hear you, Kate: in sooth you scape not so.

KATHARINA
I *chafe* you, if I tarry: let me go.

PETRUCHIO
No, not a whit: I find you passing gentle.
'Twas told me you were rough and coy and sullen,
And now I find report a very liar;
For thou are pleasant, gamesome, passing courteous,

But *slow in speech*, yet sweet as spring-time flowers:
Thou canst not frown, thou canst not look *askance*,
Nor bite the lip, as angry wenches will,
Nor hast thou pleasure to be cross in talk,
But thou with mildness entertain'st thy wooers,
With gentle conference, soft and affable.
Why does the world report that Kate doth limp?
O slanderous world! Kate like the hazel-twig
Is straight and slender and as brown in hue
As hazel nuts and sweeter than the kernels.
O, let me see thee walk: thou dost not *halt*.

KATHARINA
Go, fool, <u>and whom thou keep'st command</u>.

PETRUCHIO
<u>Did ever Dian so become a grove</u>
<u>As Kate this chamber with her princely gait?</u>
<u>O, be thou Dian, and let her be Kate;</u>
<u>And then let Kate be chaste and Dian sportful!</u>

KATHARINA
Where did you study all this goodly speech?

PETRUCHIO
It is extempore, from my *mother-wit*.

KATHARINA
<u>A witty mother! witless else her son</u>.

PETRUCHIO
Am I not wise?

KATHARINA
<u>Yes; keep you warm</u>.

PETRUCHIO
Marry, so I mean, sweet <u>Katharina</u>, in thy bed:
And therefore, setting all this chat aside,
Thus in plain terms: your father hath consented
That you shall be my wife; your dowry 'greed on;
And, Will you, nill you, I will marry you.
Now, Kate, I am a husband for your turn;
For, by this light, whereby I see thy beauty,
Thy beauty, that doth make me like thee well,
Thou must be married to no man but me;
For I am he am born to tame you Kate,
And bring you from a wild Kate to a Kate
Conformable as other household Kates.
Here comes your father: never make denial;
I must and will have <u>Katharina</u> to my wife.

Much of the scene will play—or can be made to play—as it stands, but there are key moments that risk losing the audience. If the performance is compelling, of course, the playful energy that usually drives the scene will probably keep everyone on board, but the scene works best when everyone understands at very least the gist of all the sparring, rather than merely sensing that there *is* sparring going on. As I have said before, the dramaturg's work is in detail, and while individual moments of confusion may pass without doing significant damage to the show, they can quickly accumulate to the point that they alienate the audience. Any changes made to the above scene, then, must clarify without disrupting the sense of speed and wit in the exchanges.

Let us assume that the bulk of the passage—including the italicized words—can stand, that the actors' sense of the exchange will carry the scene, and focus on those sections that need tweaking, beginning first with the parts that seem most impenetrable (those underlined and in bold).

Following from jibes about weight and seriousness, Petruchio leads Kate into a jocular taunting derived from a typically Elizabethan and proverbial use of natural history.

KATHARINA
Too light for such a *swain* as you to catch;
<u>And yet as heavy as my weight should be</u>.

PETRUCHIO
<u>Should be! should—buzz!</u>

KATHARINA
<u>Well ta'en, and like a buzzard</u>.

PETRUCHIO
<u>O slow-wing'd turtle! shall a buzzard take thee?</u>

KATHARINA
<u>Ay, for a turtle, as he takes a buzzard</u>.

PETRUCHIO
Come, come, you wasp; i' faith, you are too angry.

Figure 13.1 *The Taming of the Shrew.* GSF 2002, Dir. Garner: Gabra Zackman (Kate) and Saxon Palmer (Petruchio). Photo: Tom Meyer.

He begins by picking up her use of the word "be" and punning on "bee," using the sound the bee makes to suggest that she be quiet and listen. "Buzz," as the footnotes to most editions will tell you, was an interjection of irritation and dismissal, though it was also used to suggest rumormongering, tales being buzzed about. Kate, in turn, picks up the "buzz" sound and applies it to him; one who makes such a noise is a buzzard, a fool. Petruchio takes up the sense of himself as a buzzard (a large, slow hawk of the buteo family—not the American vulture), which can capture her, the turtle*dove*, emblem of love. Kate's answer may suggest turtle as in tortoise (withdrawing into her shell), leaving Petruchio to return to the bee joke ("Come, come, you wasp . . .").

On the page, the lines can be unraveled well enough, but such things cannot easily be explained in performance, and the use of business (or, worse, program notes) to explain difficulties of this kind raises real questions about the purpose of theatre. The lines are of no great consequence in matters of story or character since they enact a dynamic going on throughout the scene and I would recommend cutting at least the first two bolded lines rather than trying to change them into something clearer. The first two could be pulled off if the actor playing Petruchio can convey the be/bee/buzz connection, and "buzzard" can play as a generic insult, but the subsequent two lines about the turtle are not worth saving. In fact, taking out the turtle/buzzard lines makes the link from "bee" to "wasp" clearer.

The next lines similarly are part of an embedded metaphor, some of which might be salvageable though some isn't. They begin with Kate's warning that if Petruchio hits her he will not be a gentleman and will thus lose his arms in the sense of a heraldic "coat of arms":

PETRUCHIO
I swear I'll cuff you, if you strike again.

KATHARINA
<u>So may you lose your arms:</u>
<u>If you strike me, you are no gentleman;</u>
<u>And if no gentleman, why then no arms</u>.

PETRUCHIO
<u>A herald, Kate? O, put me in thy books!</u>

KATHARINA
<u>What is your crest? a coxcomb?</u>

PETRUCHIO
<u>A combless cock, so Kate will be my hen.</u>

KATHARINA
<u>No cock of mine; you crow too like a craven.</u>

Kate's line "If you strike me, you are no gentleman" is clear enough, and she seems at pains to explain her joke, but Petruchio's reference to heraldic books (with a pun on favor, as in getting into someone's "good books") is likely to be bewildering. But if the lines are cut altogether, we lose the set up for the coxcomb as crest joke, whose sexual implications are self-evident. One possible choice would be to edit the exchange thus:

PETRUCHIO
I swear I'll cuff you, if you strike again.

KATHARINA
If you strike me, you are no gentleman;
What is your crest? a coxcomb?

PETRUCHIO
A combless cock, so Kate will be my hen.

The link between the status of gentleman and a crest is now more implicit than I would like, but it is a playable transition, and his supposed foolishness (the coxcomb) is now interestingly bracketed to his threat to hit her.

The last section demanding reworking is another use of Elizabethan proverbial wisdom meaning little to a modern audience:

PETRUCHIO
It is extempore, from my mother-wit.

KATHARINA
A witty mother! witless else her son.

PETRUCHIO
Am I not wise?

KATHARINA
Yes; keep you warm.

PETRUCHIO
Marry, so I mean, sweet Katharine, in thy bed:

The proverb alluded to is "enough wit to keep oneself warm," but unless the audience knows this, Kate's line here is almost meaningless even if delivered as an attempt at a farewell. The problem with cutting the line is that we lose the set-up for the important "in thy bed" line that follows it, the most powerful statement of marital and sexual intent Petruchio has yet uttered, and one that leads into his most definitive statement of purpose. One possibility might be to replace the line with something that makes the proverb more explicit, thus:

PETRUCHIO
Am I not wise?

KATHARINA
Enough to keep you warm.

PETRUCHIO
Marry, so I mean, sweet Katharine, in thy bed:

The nature of her remark is a little clearer, and the set up for the "in thy bed" remark is retained. You will notice also that I have changed the original downloaded text's "Katharina" to "Katherine," an error that has no textual justification in early editions and that breaks the iambic pentameter of the line needlessly. I have left the earlier line "a witty mother! Witless else her son" (suggesting that what little intellect Petruchio has he inherited from his mother) because I think the actress can probably make its generally derogatory sense clear, even if the precise nature of the jab is less so. Here, as elsewhere where the lines are central to a longer exchange, which, for the most part, is fairly clear, it is better to leave the actors and director to see if they can make the lines work compellingly before removing them outright.

The same general strategy can be employed in less contentious moments:

You lie, in faith; for you are call'd plain Kate,
And bonny Kate and sometimes Kate the *curst*;
But Kate, the prettiest Kate in *Christendom*
Kate of Kate Hall, my super-dainty Kate,
For dainties are all Kates, and therefore, Kate,
Take this of me, Kate of my consolation;

The core pun of the underlined phrase—that "kates" meant "sweetmeats" or "dainties"—is one that few audience members will get, but my impulse would be to leave it since the playfulness introduces no actual confusion, even if the precise association is lost. The obscurities of the next passage are less easily playable:

PETRUCHIO
Myself am moved to woo thee for my wife.

KATHARINA
Moved! in good time: let him that moved you hither
Remove you hence: I knew you at the first
You were a moveable.

PETRUCHIO
Why, what's a moveable?

KATHARINA
A join'd-stool.

PETRUCHIO
Thou hast hit it: come, sit on me.

The first set of punning on "moved" ("myself am moved"/"remove you hence") works fine, but the link to the joint stool (punning on "moveable" meaning a piece of furniture) is less clear, as is the implication of the insult. Is she saying he's inconstant and unreliable, or merely that he is dumb as a stump? Neither is clear or especially funny. Again the difficulty with simply cutting the lines is that we need a bridge to the much clearer set of jibes about sitting and bearing. Still, this could be achieved physically, the scene working thus:

PETRUCHIO
Myself am moved to woo thee for my wife.

KATHARINA
Moved! in good time: let him that moved you hither
Remove you hence.

PETRUCHIO (*sitting*)
     Come, sit on me.

Here Petruchio's defiant response to Kate's order to leave is rendered in action (he sits down) and the action of sitting provides the segue into his invitation for her to join him. The two new half lines ("Remove you hence" and "Come, sit on me") even come close to maintaining the (not entirely consistent) metrical pattern of the scene. Likewise, in the next tricky section, it is useful to keep the meter intact because it indicates the balance of the struggle and gives it an energy that keeps it moving forward.

PETRUCHIO
Women are made to bear, and so are you.

KATHARINA
No such <u>jade</u> as you, if me you mean.

PETRUCHIO
Alas! good Kate, I will not burden thee;
For, knowing thee to be <u>but young and light</u>—

KATHARINA
Too light for such a *swain* as you to catch;
<u>And yet as heavy as my weight should be.</u>

Petruchio's use of "bear" (meaning both bear children and bear a man during intercourse) can surely be suggested by the actor (even if he has to choose only one of the pun's implications); but "jade" is trickier, it means something like "an old or worthless horse" and might be replaced by words like "nag," "mule," or "ass," all of which scan metrically being but one syllable." Their implications are a little different from "jade," of course, but they work well enough and provide the necessary link to the next lines about burdening. "Young and light" is counterintuitive for a modern audience however, since it sounds complementary, but suggests frivolousness or flightiness in the Renaissance, even lustiness. If the actress playing Kate is physically large, of course, the lines may simply be delivered as ironic (which is why the dramaturg needs to know the cast before he prepares the script), but otherwise they need adjusting. Kate's reiteration of "light" in the next line suggests speed or elusiveness, an escapability beyond the means of such a "swain" (country bumpkin lover) as Petruchio, though she also claims a weightiness (seriousness, soberness) appropriate to her standing. An alternate version may be crafted thus:

PETRUCHIO
Women are made to bear, and so are you.

KATHARINA
No such ass as you, if me you mean.

PETRUCHIO
Alas! good Kate, I will not burden thee;
For, knowing thee to be but frail and fast—
KATHARINA
Too fast for such a clown as you to catch;
And yet as serious as I should be.

The change of "young and light" to "frail and fast" obviously changes the line and its meaning, but it retains some of the original's implications (frail = flightiness, lack of strength; fast = sexually aggressive, which she recrafts simply as speed in the next line) while utilizing words common throughout Shakespeare and introducing an appealing alliteration in the process. Most audiences would grasp the meaning as well as the tone better than in the original, and only those who know the scene thoroughly will notice the change. "Swain" does not easily translate (particularly into a single syllable word, which is what is required to maintain the meter); "clown" loses the romantic sense associated ironically with the original word, though it is, again, a familiar Shakespearean term often used of low, rustic idiots, and resonates better with a modern audience. In removing the "light" metaphor, I had to adjust that of "heavy" in the last line, and though "serious" is less clearly linked to the image of speed in the previous lines, it conveys her point, retains the original meter, and sets up the "should be, should buzz" of the next line well enough.

The following lines hinge on the use of the word "crab" to mean "crabapple":

PETRUCHIO
Nay, come, Kate, come; you must not look so sour.
KATHARINA
<u>It is my fashion, when I see a **crab**</u>.
PETRUCHIO
<u>Why, here's no **crab**; and therefore look not sour</u>.
KATHARINA
There is, there is.

"Crab" delivered like this without explanation suggests to a modern audience the crustacean, not the fruit, and that is perfectly playable, though crabs are less sour than they are—potentially—repulsive or scary. Addition of the word "apple" to "crab" plays havoc with the meter and disrupts the speed of the exchange, so the lines either have to stand (with their crustacean over-tones), be cut outright, or more radically rewritten, all of which are acceptable since the lines are not especially significant or amusing of themselves. One adaptive strategy might be to rethink the source of the sourness, though the number of single syllable nouns clearly linked to sourness is small and largely unsuitable. It should be said, however, that the original's link between looking sour and *seeing*, not tasting, a crab is pretty thin, an objection that might justify alternatives. "Crab" could be replaced by some other noun at which one may pull a face, like a "skunk," for example (though skunks don't populate the English Renaissance), shifting the issue of tasting to smelling, though such a substitution is perhaps conspicuous enough to seem invasive. "Lemon" might work, and though it adds a second syllable it fits both the required sourness and the general sense (albeit modern and usually applied to cars) of insult as Kate then uses it.

The use-value of the following lines may hinge on the ages of the actors who play them. The exchange is not especially opaque, but neither is it especially funny or witty, and I would consider eliminating it.

> **PETRUCHIO**
> What, you mean my face?
>
> **KATHARINA**
> Well aim'd of such a young one.
>
> **PETRUCHIO**
> Now, by Saint George, I am too young for you.
>
> **KATHARINA**
> Yet you are wither'd.
>
> **PETRUCHIO**
> 'Tis with *cares*.
>
> **KATHARINA**
> I care not.

Kate's first use of "young" suggests "inexperienced" or perhaps more simply "clueless," but Petruchio's subsequent use suggests "strong," neither of which are instantly apparent to the modern audience, for whom the exchange goes by too fast for much consideration. It might be a shame to lose the final play on "care," however, so one option would be to condense the passage:

> **PETRUCHIO**
> What, you mean my face? 'Tis withered o'er with cares.
>
> **KATHARINA**
> I care not.

The extra half line or so (the line is not quite metrical but neither is the passage it replaces) maintains Petruchio's self-deprecation (and does so in more mock Shakespearean terms that few people will spot as interpolated), omits the pun on "young" and moves the encounter forward quickly.

A little later in the scene, Petruchio attests to the way that Kate belies what he has heard about her and she replies, "Go, fool, *and whom thou keep'st command*." Her response (to his telling her to walk about so he can admire her gait[1]), tells him only to give orders to those who work for him, but the syntax is potentially confusing for a modern audience and it might be worth rephrasing. One way would be to simply rearrange the remark: "Go, fool, command those whom thou keep'st." This is fractionally clearer, I think, because the verb comes earlier, but a more adaptive rewriting might say

"Go, fool, command your servants only." This is clearer still and, like the other version, is close enough to the meter of the original that the alteration does not jar on the ear unduly.

Petruchio's response demands some sense of Greco-Roman mythology to be absolutely clear:

PETRUCHIO
<u>Did ever Dian so become a grove</u>
<u>As Kate this chamber with her princely gait?</u>
<u>O, be thou Dian, and let her be Kate;</u>
<u>And then let Kate be chaste and Dian sportful!</u>
KATHARINA
Where did you study all this goodly speech?

It would be easy to remove Petruchio's four lines (tacking Kate's subsequent question onto her previous remark, or using it in place of that remark), but Petruchio is being deliberately flowery in his invocation of the goddess of chastity (as spied, perhaps, by Actaeon), as Kate's remark about his "goodly speech" makes clear. As such, I think that most audience members—even if they didn't get the reference directly—would roll with the sense that this is hyperbolic poetizing, the precise meaning of which is less important than its tone. I would leave it as is in a first pass, being prepared to revisit it if the script needed further cutting for running time, or if the actors couldn't find a way to make it work.

EXAMPLE 2: EXPOSITION: *THE COMEDY OF ERRORS*, 1, 1

In this example, rather than including a lengthy discussion of each possible choice, I have simply presented the edited script with the original. The potentially problematic words or phrases in the original are italicized. The edited choices in the second version, all of which are minor, are marked in bold. All such choices have been made for clarity and, secondarily, to abbreviate the lengthy exposition that can so take the life out of the production. Any change needing further clarification has been footnoted. This is broadly the kind of format I use to mark cuts in the text for review by the director (Sabin Epstein, who directed the show for GSF in 1999).

Original Version

Act 1, Scene 1
A hall in DUKE SOLINUS'S palace.
Enter DUKE SOLINUS, EGEON, Gaoler,
Officers, and other Attendants
EGEON
  Proceed, Solinus, to *procure* my fall
  And by the doom of death end woes and all.

DUKE SOLINUS

> Merchant of Syracusa, plead no more;
> I am not partial to infringe our laws:
> The enmity and discord which of late
> Sprung from the rancorous outrage of your duke
> To merchants, our well-dealing countrymen,
> Who wanting *guilders* to redeem their lives
> Have seal'd his rigorous statutes with their bloods,
> Excludes all pity from our threatening looks.
> For, since the mortal and *intestine jars*
> 'Twixt thy seditious countrymen and us,
> It hath in solemn synods been decreed
> *Both by the Syracusians and ourselves,*
> *To admit no traffic to our hostile towns. Nay, more,*
> If any born at Ephesus be seen
> At any Syracusian marts and fairs;
> Again: if any Syracusian born
> Come to the bay of Ephesus, he dies,
> His goods confiscate to the duke's dispose,
> Unless a thousand marks be leviéd,
> To quit the penalty and to ransom him.
> Thy substance, valued at the highest rate,
> Cannot amount unto a hundred marks;
> Therefore by law thou art condemned to die.

EGEON

> Yet this my comfort: when your words are done,
> My woes end likewise with the evening sun.

DUKE SOLINUS

> Well, Syracusian, say in brief the cause
> Why thou departed'st from thy native home
> And for what cause thou camest to Ephesus.

EGEON

> A heavier task could not have been imposed
> Than I to speak my griefs unspeakable:
> *Yet, that the world may witness that my end*
> *Was wrought by nature, not by vile offence,*
> *I'll utter what my sorrows give me leave.*
> In Syracusa was I born, and wed
> Unto a woman, happy but for me,
> *And by me, had not our luck been bad.*
> With her I lived in joy; our wealth increased
> By prosperous voyages I often made
> To Epidamnum; *till my factor's death*
> *And the great care of goods at random left*
> *Drew me from kind embracements of my spouse:*
> *From whom my absence was not six months old*
> *Before herself, (almost at fainting under*
> *The pleasing punishment that women bear),*
> *Had made provision for her following me*
> *And soon and safe arrived where I was.*

There *had she* not been long, but she became
A joyful mother of two goodly sons;
And, which was strange, the one so like the other,
As could not be distinguish'd but by names.
That very hour, and in the self-same inn,
A *meaner* woman was delivered
Of such a burden, male twins, both alike:
Those,—for their parents were exceeding poor,—
I bought and brought up to attend my sons.
My wife, *not meanly* proud of *two such* boys,
Made daily motions for our home return:
Unwilling I agreed. Alas! too soon,
We came aboard.
A league from Epidamnum had we sail'd,
Before the always wind-obeying deep
Gave any tragic *instance* of our harm:
But longer did we not retain much hope;
For what obscured light the heavens did grant
Did but convey unto our fearful minds
A *doubtful* warrant of immediate death;
Which though myself would gladly have embraced,
Yet the incessant weepings of my wife,
Weeping before for what she saw must come,
And piteous plainings of the pretty babes,
That mourn'd for custom, ignorant what to fear,
Forced me to seek delays for them and me.
And this it was, for other means was none:
The sailors sought for safety by our boat,
And left the ship, then sinking-ripe, to us:
My wife, more careful for the latter-born,
Had fasten'd him unto a small spare mast,
Such as seafaring men provide for storms;
To him one of the other twins was bound,
Whilst I had been like heedful of the other:
The children thus disposed, my wife and I,
Fixing our eyes on whom our care was fix'd,
Fasten'd ourselves at either end the mast;
And floating straight, obedient to the stream,
Was carried towards Corinth, as we thought.
At length the sun, gazing upon the earth,
Dispersed those *vapors* that offended us;
And by the benefit of his wished light,
The seas wax'd calm, and we discovered
Two ships from far making amain to us,
Of Corinth that, of Epidaurus this:
But ere they came,—O, let me say no more!
Gather the sequel by *that* went before.

DUKE SOLINUS
   Nay, forward, old man; do not break off so;
   For we may pity, though not pardon thee.

EGEON

> O, had the gods done so, I had not now
> Deservedly term'd them merciless to us!
> For, ere the ships could meet by twice five leagues,
> We were encountered by a mighty rock;
> Which being violently borne upon,
> Our helpful *ship* was splitted in the midst;
> *So that, in this unjust divorce of us,*
> *Fortune had left to both of us alike*
> *What to delight in, what to sorrow for.*
> *Her part, poor soul! seeming as burdened*
> *With lesser weight but not with lesser woe,*
> Was carried with more speed before the wind;
> And in our sight they three were taken up
> By fishermen of Corinth, as we thought.
> At length, another *ship had* seized on us;
> And, knowing whom it was their *hap* to save,
> Gave healthful welcome to their shipwreck'd guests;
> And would have robbed the fishers of their prey,
> Had not their *bark* been very slow of sail;
> And therefore homeward did they bend their course.
> Thus have you heard me sever'd from my bliss;
> That by misfortunes was my life prolong'd,
> To tell sad stories of my own mishaps.

DUKE SOLINUS

> And for the sake of them thou sorrowest for,
> Do me the favour to relate at full
> What hath befall'n of them and thee till now.

EGEON

> My *youngest* boy, and yet my *eldest* care,
> At eighteen years became inquisitive
> After his brother: and importuned me
> That his attendant, so his case was like,
> Reft of his brother but retain'd his name,
> Might bear him company in the quest of him:
> *Whom whilst I labour'd of a love to see,*
> *I hazarded the loss of whom I loved.*
> Five summers have I spent in furthest Greece,
> Roaming clean through the bounds of Asia,
> And, coasting homeward, came to Ephesus;
> Hopeless to find, yet loath to leave unsought
> *Or* that or any place that harbours men.
> But here must end the story of my life;
> And happy were I in my timely death,
> Could all my travels warrant me they live.

DUKE SOLINUS

> Hapless Egeon, whom the fates have mark'd
> To bear the extremity of dire mishap!
> Now, trust me, were it not against our laws,

Against my crown, against my office,
My soul would sue as advocate for thee.
But, though thou art adjudged to the death
And passed sentence may not be recall'd
But to our honour's great disparagement,
Yet I will favour thee in what I can.
Therefore, merchant, I'll limit thee this day
To seek thy life by beneficial help:
Try all the friends thou hast in Ephesus;
Beg thou, or borrow, to make up the sum,
And live; if no, then thou art doom'd to die.
Gaoler, take him to thy custody.

GAOLER

I will, my lord.

EGEON

Hopeless and helpless doth Egeon wend,
But to procrastinate his lifeless end.
Exeunt.

Edited Version

Act 1, Scene 1
A hall in DUKE SOLINUS'S palace.
Enter DUKE SOLINUS, EGEON, Gaoler,
Officers, and other Attendants

EGEON

Proceed, Solinus, to **pronounce** my fall
And by the doom of death end woes and all.

DUKE SOLINUS

Merchant of Syracusa, plead no more;
I am not partial to infringe our laws:
The enmity and discord which of late
Sprung from the rancorous outrage of your duke
To merchants, our well-dealing countrymen,
Who wanting **moneys** to redeem their lives
Have seal'd his rigorous statutes with their bloods,
Excludes all pity from our threatening looks.
For, since the mortal and **internal wars**
'Twixt thy seditious countrymen and us,
It hath in solemn synods been decreed
Both by the Syracusians and ourselves,
To admit no traffic to our hostile towns. Nay, more,[2]
If any born at Ephesus be seen
At any Syracusian marts and fairs;
Again: if any Syracusian born
Come to the bay of Ephesus, he dies,
His goods confiscate to the duke's dispose,
Unless a thousand marks be leviéd,
To quit the penalty and to ransom him.

Thy substance, valued at the highest rate,
Cannot amount unto a hundred marks;
Therefore by law thou art condemned to die.

EGEON

Yet this my comfort: when your words are done,
My woes end likewise with the evening sun.

DUKE SOLINUS

Well, Syracusian, say in brief the cause
Why thou departed'st from thy native home
And for what cause thou camest to Ephesus.

EGEON

A heavier task could not have been imposed
Than I to speak my griefs unspeakable:
[CUT Yet, that the world may witness that my end
Was wrought by nature, not by vile offence,
I'll utter what my sorrows give me leave.]
In Syracusa was I born, and wed
Unto a woman, happy but for me,
[CUT And by me, had not our luck been bad.]
With her I lived in joy; our wealth increased
By prosperous voyages I often made
To Epidamnum; [CUT till my factor's death³
And the great care of goods at random left
Drew me from kind embracements of my spouse:
From whom my absence was not six months old
Before herself, (almost at fainting under
The pleasing punishment that women bear),
Had made provision for her following me
And soon and safe arrived where I was.]
There **we'd** not been long, but she became
A joyful mother of two goodly sons;
And, which was strange, the one so like the other,
As could not be distinguish'd but by names.
That very hour, and in the self-same inn,
A **serving**⁴ woman was delivered
Of such a burden, male twins, both alike:
Those,—for their parents were exceeding poor,—
I bought and brought up to attend my sons.
My wife, **extremely** proud of **our two** boys,
Made daily motions for our home return:
Unwilling I agreed. Alas! too soon,
We came aboard.
A league from Epidamnum had we sail'd,
Before the always wind-obeying deep
Gave any tragic **warning** of our harm:
But longer did we not retain much hope;
For what obscured light the heavens did grant
Did but convey unto our fearful minds
A **dreadful** warrant of immediate death;

[CUT **Which though myself would gladly have embraced,**
Yet the incessant weepings of my wife,
Weeping before for what she saw must come,
And piteous plainings of the pretty babes,
That mourn'd for custom, ignorant what to fear,
Forced me to seek delays for them and me.
And this it was, for other means was none:][5]
The sailors sought for safety by our boat,
And left the ship, then sinking-ripe, to us:
My wife, more careful for the latter-born,
Had fasten'd him unto a small spare mast,
Such as seafaring men provide for storms;
To him one of the other twins was bound,
Whilst I had been like heedful of the other:
The children thus disposed, my wife and I,
Fixing our eyes on whom our care was fix'd,
Fasten'd ourselves at either end the mast;
And floating straight, obedient to the stream,
Were carried towards Corinth, as we thought.
At length the sun, gazing upon the earth,
Dispersed those **storm clouds** that offended us;
And by the benefit of his wished light,
The seas wax'd calm, and we discovered
Two ships from far making amain to us,
Of Corinth that, of Epidaurus this:
But ere they came,—O, let me say no more!
Gather the sequel by **what** went before.

DUKE SOLINUS

Nay, forward, old man; do not break off so;
For we may pity, though not pardon thee.

EGEON

O, had the gods done so, I had not now
Deservedly term'd them merciless to us!
For, ere the ships could meet by twice five leagues,
We were encountered by a mighty rock;
Which being violently borne upon,
Our helpful **mast** was splitted in the midst;[6]
So that, in this unjust divorce of us,
Fortune had left to both of us alike
What to delight in, what to sorrow for.
Her part, poor soul! seeming as burdened
With lesser weight but not with lesser woe,[7]
Was carried with more speed before the wind;
And in our sight they three were taken up
By fishermen of Corinth, as we thought.
At length, another **vessel** seized on us;[8]
And, knowing whom it was their **luck** to save,
Gave healthful welcome to their shipwreck'd guests;
And would have robbed the fishers of their prey,
Had not their **ship** been very slow of sail;

And therefore homeward did they bend their course.
Thus have you heard me sever'd from my bliss;
That by misfortunes was my life prolong'd,
To tell sad stories of my own mishaps.

DUKE SOLINUS

And for the sake of them thou sorrowest for,
Do me the favour to relate at full
What hath befall'n of them and thee till now.

EGEON

My **eldest** boy, and yet my **youngest** care,[9]
At eighteen years became inquisitive
After his brother: and importuned me
That his attendant, so his case was like,
Reft of his brother but retain'd his name,
Might bear him company in the quest of him:
And since that time I have not seen my son
Nor heard aught of his brother whom he sought.[10]
Five summers have I spent in furthest Greece,
Roaming clean through the bounds of Asia,
And, coasting homeward, came to Ephesus;
Hopeless to find, yet loath to leave unsought
In that or any place that harbours men.
But here must end the story of my life;
And happy were I in my timely death,
Could all my travels warrant me they live.

DUKE SOLINUS

Hapless Egeon, whom the fates have mark'd
To bear the extremity of dire mishap!
Now, trust me, were it not against our laws,
Against my crown, against my office,
My soul would sue as advocate for thee.
But, though thou art adjudged to the death
And passed sentence may not be recall'd
But to our honour's great disparagement,
Yet I will favour thee in what I can.
Therefore, merchant, I'll limit thee this day
To seek thy life by beneficial help:
Try all the friends thou hast in Ephesus;
Beg thou, or borrow, to make up the sum,
And live; if no, then thou art doom'd to die.
Gaoler, take him to thy custody.

GAOLER

I will, my lord.

EGEON

Hopeless and helpless doth Egeon wend,
But to procrastinate his lifeless end.
Exeunt.

Example 3: Heavier Cutting: *Julius Caesar* 2.1

The following cutting represents a script for a 2001 production that the director, John Dillon, wanted to move swiftly toward the assassination of Caesar without overly reducing the range and complexity of the central characters. Except in a few matters of sentence-level clarity, the editorial agenda was thus to shorten where possible, and the director came in with some of the larger cuts already in mind. Since the production was set in 1930s Louisiana, with Caesar modeled loosely on Huey Long, we particularly looked to cut or reduce those moments in which the text went out of its way to emphasize the Romanness or general Classicism of the play. Changes from the copy text have been marked in bold with explanatory footnotes added. In some places I have marked more tentative cutting suggestions with a question mark.

ACT II
SCENE I. Rome. BRUTUS's orchard.
Enter BRUTUS

BRUTUS
What, Lucius, ho!
I cannot, by the progress of the stars,
Give guess how near to day. Lucius, I say!
I would it were my fault to sleep so soundly.
[CUT When, Lucius, when?][11] Awake, I say! what, Lucius!

Enter LUCIUS

LUCIUS
Call'd you, my lord?

BRUTUS
Get me a taper in my study, Lucius:
When it is lighted, come and call me here.

LUCIUS
I will, my lord.
Exit

BRUTUS
It must be by his death: and for my part,
I know no personal cause to spurn at him,
But for the general. He would be crown'd:
How that might change his nature, there's the question.
It is the bright day that brings forth the **viper**;[12]
And that craves wary walking. Crown him **king**;[13]
And then, I grant, we put a sting in him,
That at his will he may do danger with.
The abuse of greatness is, when it disjoins
Remorse from power: **Though**[14] to speak truth of Caesar,
I have not known when his **desires**[15] sway'd
More than his reason. But 'tis a common proof,

That lowliness is young ambition's ladder,
Whereto the climber-upward turns his face;
But when he once attains the upmost round.
He then unto the ladder turns his back,
Looks in the clouds, scorning the base degrees
By which he did ascend. So Caesar may.
Then, lest he may, prevent. And, since the quarrel
Cannot seem just because of what he is,[16]
Fashion it thus; that what he is, augmented,
Would run to these and these extremities:
And therefore think him as a serpent's egg
Which, hatch'd, would, as his kind, grow mischievous,
And kill him in the shell.

Re-enter LUCIUS

LUCIUS

The taper burneth in your closet, sir.
Searching the window for a flint, I found
This paper, thus seal'd up; and, I am sure,
It did not lie there when I went to bed.
Gives him the letter

BRUTUS

Get you to bed again; it is not day.
Is not to-morrow, boy, the **Ides** of March?[17]

LUCIUS

I know not, sir.

BRUTUS

Look in the calendar, and bring me word.

LUCIUS

I will, sir.
Exit

BRUTUS

The exhalations whizzing in the air
Give so much light that I may read by them.
Opens the letter and reads
'Brutus, thou sleep'st: awake, and see thyself.
Shall Rome, etc. Speak, strike, redress!
Brutus, thou sleep'st: awake!'
Such instigations have been often dropp'd
Where I have took them up.
"Shall Rome, etc." Thus must I piece it out:
Shall Rome stand under one man's awe? What, Rome?
My ancestors did from the streets of Rome
The Tarquin drive, when he was call'd a king.
"Speak, strike, redress!" Am I entreated
To speak and strike? O Rome, I make thee promise:
If the redress will follow, thou receivest
Thy full petition at the hand of Brutus!
Re-enter LUCIUS

LUCIUS
Sir, March is wasted **fourteen** days.[18]
Knocking within

BRUTUS
'Tis good. Go to the gate; somebody knocks.
Exit LUCIUS
Since Cassius first did whet me against Caesar,
I have not slept.
Between the acting of a dreadful thing
And the first motion, all the interim is
Like a phantasma, or a hideous dream:
The **spirit**[19] and **the baser instincts**[20]
Are then in council; and the state of man,
Like to a little kingdom, suffers then
The nature of an insurrection.
Re-enter LUCIUS

LUCIUS
Sir, 'tis [**CUT your brother**][21] Cassius at the door,
Who doth desire to see you.

BRUTUS
Is he alone?

LUCIUS
No, sir, there are more with him.

BRUTUS
Do you know them?

LUCIUS
No, sir; their hats are pluck'd about their ears,
And half their faces buried in their cloaks,
That by no means I may discover them
[**CUT? By any mark of favor.**][22]

BRUTUS
Let 'em enter.
Exit LUCIUS
They are the faction. O conspiracy,
Shamest thou to show thy dangerous brow by night,
When evils are most free? O, then by day
Where wilt thou find a cavern dark enough
To mask thy monstrous visage? Seek none, conspiracy;
Hide it in smiles and affability:
[**CUT: For if thou put, thy native semblance on,
Not Erebus itself were dim enough
To hide thee from prevention.**][23]
*Enter the conspirators, CASSIUS, CASCA, DECIUS, CINNA, METELLUS,
and TREBONIUS*

CASSIUS
I think we are too bold upon your rest:
Good morrow, Brutus; do we trouble you?

BRUTUS
I have been up this hour, awake all night.
Know I these men that come along with you?

CASSIUS
Yes, every man of them, and no man here
But honours you; and every one doth wish
You had but that opinion of yourself
Which every noble Roman bears of you.
This is Trebonius.

BRUTUS
He is welcome hither.

CASSIUS
This, Decius Brutus.

BRUTUS
He is welcome too.

CASSIUS
This, Casca; this, Cinna; and this, Metellus Cimber.

BRUTUS
They are all welcome.
[CUT **What watchful cares do interpose themselves**
Betwixt your eyes and night?

CASSIUS
Shall I entreat a word?
BRUTUS and CASSIUS whisper

DECIUS BRUTUS
Here lies the east: doth not the day break here?

CASCA
No.

CINNA
O, pardon, sir, it doth; and yon gray lines
That fret the clouds are messengers of day.

CASCA
You shall confess that you are both deceived.
Here, as I point my sword, the sun arises,
Which is a great way growing on the south,
Weighing the youthful season of the year.
Some two months hence up higher toward the north
He first presents his fire; and the high east
Stands, as the Capitol, directly here.][24]

BRUTUS
Give me your hands all over, one by one.

CASSIUS
And let us swear our resolution.

BRUTUS
No, not an oath: [CUT **if not the face of men,**
The sufferance of our souls, the time's abuse,—
If these be motives weak, break off betimes,

And every man hence to his idle bed;
So let high-sighted tyranny range on,
Till each man drop by lottery. But][25] if these,
As I am sure they do, bear fire enough
To kindle cowards and to steel with valour
The melting spirits of women, then, countrymen,
What need we any spur but our own cause,
To prick us to redress? what other bond
Than secret Romans, that have spoke the word,
And will not **falter?**[26] and what other oath
Than honesty to honesty engaged,
That this shall be, or we will fall for it?
[CUT Swear priests and cowards and men cautelous,
Old feeble carrions and such suffering souls
That welcome wrongs; unto bad causes swear
Such creatures as men doubt; but][27] **D**o not stain
The even virtue of our enterprise,
Nor the insuppressive mettle of our spirits,
To think **[CUT that or]** our cause or our performance
Did need an oath; when every drop of blood
That every Roman bears, and nobly bears,
Is guilty of a several bastardy,
If he do break the smallest particle
Of any promise that hath pass'd from him.

CASSIUS
But what of Cicero? shall we sound him?
I think he will stand very strong with us.

CASCA
Let us not leave him out.

CINNA
No, by no means.

METELLUS CIMBER
O, let us have him, for his silver hairs
Will purchase us a good opinion
And buy men's voices to commend our deeds:
It shall be said, his judgment ruled our hands;
Our youths and wildness shall no whit appear,
But all be buried in his gravity.

BRUTUS
O, name him not: let us not break with him;
For he will never follow any thing
That other men begin.

CASSIUS
Then leave him out.

CASCA
Indeed he is not fit.

DECIUS BRUTUS
Shall no man else be touch'd but only Caesar?

CASSIUS
Decius, well urged: I think it is not **fit**,[28]
Mark Antony, so well beloved of Caesar,
Should outlive Caesar: we shall find of him
A shrewd contriver; and, you know, his means,
If he improve them, may well stretch so far
As to **confound**[29] us all: which to prevent,
Let Antony and Caesar fall together.

BRUTUS
Our course will seem too bloody, Caius Cassius,
To cut the head off and then hack the limbs,
[CUT?: Like wrath in death and envy afterwards;]
For Antony is but a limb of Caesar:
Let us be sacrificers, but not butchers, Caius.
We all stand up against the spirit of Caesar;
And in the spirit of men there is no blood:
O, that we then could come by Caesar's spirit,
And not dismember Caesar! But, alas,
Caesar must bleed for it! **So**,[30] gentle friends,
Let's kill him boldly, but not wrathfully;
Let's carve him as a dish fit for the gods,
Not hew him as a carcass fit for hounds:
**[CUT?: And let our hearts, as subtle masters do,
Stir up their servants to an act of rage,
And after seem to chide 'em.]** This shall make
Our purpose necessary and not envious:
Which so appearing to the common eyes,
We shall be call'd purgers, not murderers.
And for Mark Antony, think not of him;
For he can do no more than Caesar's arm
When Caesar's head is off.

CASSIUS
Yet I fear him;
For in the engrafted love he bears to Caesar—

BRUTUS
Alas, good Cassius, do not think of him:

**[CUT?: If he love Caesar, all that he can do
Is to himself, take thought and die for Caesar:
And that were much he should;]**[31] for he is given
To sports, to wildness and much company.

TREBONIUS
There is no **danger**[32] in him; let him not die;
For he will live, and laugh at this hereafter.
Clock strikes

BRUTUS
Peace! count the clock.

CASSIUS
The clock hath stricken three.

TREBONIUS
'Tis time to part.

CASSIUS
But it is doubtful yet,
Whether Caesar will come forth to-day, or no;
For he is superstitious grown of late,
Quite from the main opinion he held once
Of fantasy, of dreams and ceremonies:
It may be, these apparent prodigies,
The unaccustom'd terror of this night,
And the persuasion of his **sooth-sayers**,[33]
May hold him from the Capitol to-day.

DECIUS BRUTUS
Never fear that: if he be so resolved,
I can o'ersway him; for he loves to hear
That [CUT: **unicorns may be betray'd with trees,
And bears with glasses, elephants with holes,
Lions with toils and**] men may be betrayed by[34] flatterers;
But when I tell him he hates flatterers,
He says he does, being then most flattered.
Let me work;
[CUT: **For I can give his humor the true bent,**]
And I will bring him to the Capitol.

CASSIUS
Nay, we will all of us be there to fetch him.

[CUT: **BRUTUS
By the eighth hour: is that the uttermost?**

**CINNA
Be that the uttermost, and fail not then.**

**METELLUS CIMBER
Caius Ligarius doth bear Caesar hard,
Who rated him for speaking well of Pompey:
I wonder none of you have thought of him.**

**BRUTUS
Now, good Metellus, go along by him:
He loves me well, and I have given him reasons;
Send him but hither, and I'll fashion him.**][35]

CASSIUS
The morning comes upon 's: we'll leave you, Brutus.
And, friends, disperse yourselves; but all remember
What you have said, and show yourselves true Romans.

BRUTUS
Good gentlemen, look fresh and merrily;
Let not our looks put on our purposes,
But bear it as our Roman actors do,

With untired spirits and formal constancy:
And so good morrow to you every one.
Exeunt all but BRUTUS
[CUT?: Boy! Lucius! Fast asleep? It is no matter;
Enjoy the honey-heavy dew of slumber:
Thou hast no figures nor no fantasies,
Which busy care draws in the brains of men;
Therefore thou sleep'st so sound.][36]
Enter PORTIA

PORTIA
Brutus, my lord!

BRUTUS
Portia, what mean you? wherefore rise you now?
It is not for your health thus to commit
Your weak condition to the raw cold morning.[37]

PORTIA
Nor for yours neither. You've **secretly,**[38] Brutus,
Stolen from my bed: and yesternight, at supper,
You suddenly arose, and walk'd about,
Musing and sighing, with your arms across,
And when I ask'd you what the matter was,
You stared upon me with ungentle looks;
I urged you further; then you scratch'd your head,
And too impatiently stamp'd with your foot;
Yet I insisted, yet you answer'd not,
But, with an angry wafture of your hand,
Gave sign for me to leave you: so I did;
Fearing to strengthen that impatience
Which seem'd too much enkindled, and withal
Hoping it was but an effect of **mood,**[39]
Which sometime hath his hour with every man.
It will not let you eat, nor talk, nor sleep,
And could it work so much upon your shape
As it hath much prevail'd on your condition,
I should not know you, Brutus. Dear my lord,
Make me acquainted with your cause of grief.

BRUTUS
I am not well in health, and that is all.

PORTIA
Brutus is wise, and, were he not in health,
He would embrace the means to come by it.

BRUTUS
Why, so I do. Good Portia, go to bed.

PORTIA
Is Brutus sick? and is it **healthy then**[40]
To walk unbraced and **breathe the air**[41]
Of the dank morning? **[CUT?: What, is Brutus sick,**
And will he steal out of his wholesome bed,

To dare the vile contagion of the night
And tempt the rheumy and unpurged air
To add unto his sickness?][42] No, my Brutus;
You have some sick offence within your mind,
Which, by the right and virtue of my place,
I ought to know of: [CUT: and, upon my knees,][43]
I charm you, by my once-commended beauty,
By all your vows of love and that great vow
Which did incorporate and make us one,
That you unfold to me, yourself, your half,
Why you are heavy, and what men to-night
Have had resort to you: for here have been
Some six or seven, who did hide their faces
Even from darkness.

BRUTUS
Ask not, gentle Portia.[44]

PORTIA
I should not need, if you were gentle Brutus.
Within the bond of marriage, tell me, Brutus,
Is it excepted I should know no secrets
That appertain to you? Am I yourself
But, as it were, in sort or limitation,
To keep with you at meals, comfort your bed,
And talk to you sometimes? Dwell I but in the suburbs
Of your good pleasure? If it be no more,
Portia is Brutus' harlot, not his wife.

BRUTUS
You are my true and honourable wife,
As dear to me as are the ruddy drops
That visit my sad heart.

PORTIA
[CUT: If this were true, then should I know this secret.
I grant I am a woman; but withal
A woman that Lord Brutus took to wife:
I grant I am a woman; but withal
A woman well-reputed, Cato's daughter.
Think you I am no stronger than my sex,
Being so father'd and so husbanded?][45]
Tell me your counsels, I will not disclose 'em:
[CUT I have made strong proof of my constancy,
Giving myself a voluntary wound
Here, in the thigh: can I bear that with patience.
And not my husband's secrets?][46]

BRUTUS
O ye gods,
Render me worthy of this noble wife!
Knocking within
Hark, hark! one knocks: Portia, go in awhile;
And by and by thy bosom shall partake

The secrets of my heart.
All my engagements I will construe to thee,
All the **significance**[47] of my sad brows.
[CUT Leave me with haste.[48]

Exit PORTIA
Lucius, who's that knocks?
Re-enter LUCIUS with LIGARIUS

LUCIUS
He is a sick man that would speak with you.

BRUTUS
Caius Ligarius, that Metellus spake of.
Boy, stand aside. Caius Ligarius! how?

LIGARIUS
Vouchsafe good morrow from a feeble tongue.

BRUTUS
O, what a time have you chose out, brave Caius,
To wear a kerchief! Would you were not sick!

LIGARIUS
I am not sick, if Brutus have in hand
Any exploit worthy the name of honour.

BRUTUS
Such an exploit have I in hand, Ligarius,
Had you a healthful ear to hear of it.

LIGARIUS
By all the gods that Romans bow before,
I here discard my sickness! Soul of Rome!
Brave son, derived from honourable loins!
Thou, like an exorcist, hast conjured up
My mortified spirit. Now bid me run,
And I will strive with things impossible;
Yea, get the better of them. What's to do?

BRUTUS
A piece of work that will make sick men whole.

LIGARIUS
But are not some whole that we must make sick?

BRUTUS
That must we also. What it is, my Caius,
I shall unfold to thee, as we are going
To whom it must be done.

LIGARIUS
Set on your foot,
And with a heart new-fired I follow you,
To do I know not what: but it sufficeth
That Brutus leads me on.

BRUTUS
Follow me, then.]
Exeunt

Example 4: Selective Contemporary Reference: Song Usage in *Twelfth Night* 2.3

The following brief extract deals with pieces from songs known to the Renaissance audience.[49] While there certainly are ways such fragments could be made semantically clear and funny on stage today, the songs' former familiarity—and thus, their ironic value in the scene—gets lost in modern productions. Moreover, the use of the old songs tends usually to layer the scene with a markedly unfunny dust, muting the energy of the moment. I generally advocate caution in inserting self-consciously modern references into the scripts of Renaissance plays for fear of creating what audience members often feel to be a temporally discordant note—something that occurs even in modern dress or eclectic productions, where the non-textual elements of the show are accepted as framing the details of the production's world. These old songs, however, are a special case. They are elements of the original productions' culture rendered textually, and as such they are gestures to a world beyond the confines of the play, the world of the spectators. The inverted commas between which such song fragments often appear on the page of the Shakespeare text is indicative of how they were heard: as quotation, as shared cultural memory, or as parody. Modernizing the songs not only brings the spirit of the moment firmly into the present, it does so without straying from those inverted commas, thus dodging the question of tinkering with the more clearly Shakespearean text (the dialogue) and reaffirming the play's gesture beyond itself to the larger culture beyond the theatre.

Having said that, the selection of new songs should be made with an eye on the period and setting of the production (a show that is otherwise set in the eighteenth century needs a good reason to start singing snatches of Beatles songs), and to the tone of the moment. In this latter case, of course, comedy is easier to adjust than tragedy, though I think there is real mileage to be got out of the application of this modernizing principle to those songs that appear as quotation or as proverb in the tragedies, as has been demonstrated, for example, in recent productions of *King Lear*.[50] The range of possibility is, of course, close to infinite, and the examples cited below are arbitrary choices designed merely to illustrate the point. The dramaturg who makes such suggestions should go into rehearsal armed with alternate choices with which the actors and director can experiment. In this, as in many things, one of the dramaturg's best friends is a good internet search engine; there are plenty of sites storing the lyrics to thousands of songs and these can be especially useful if the dramaturg wants to connect the song chosen to a word in the surrounding dialogue, as happens in the example below. I am assuming a contemporary—or temporally eclectic—setting for the production.

The following lines also contain other cuts and changes, which are footnoted. The sung lines are italicized.

MARIA

> What a caterwauling do you keep here! If my lady
> have not called up her steward Malvolio and bid him
> turn you out of doors, never trust me.

SIR TOBY BELCH

> "*We three kings of Orient are!*"[51] Am not
> I consanguineous? am I not of her blood?
> Tilly-vally. Lady!
> "*She's a lady! Woah, woah, woah- she's a lady!*"[52]

CLOWN

> Beshrew me, the knight's in admirable fooling.

SIR ANDREW

> Ay, he does well enough if he be disposed, and so do
> I too: he does it with a better grace, but I do it
> more natural.

SIR TOBY BELCH

> "*On the twelfth day of Christmas,*"—[53]

MARIA

> For the love o' God, peace!
> [Enter MALVOLIO]

MALVOLIO

> My masters, are you mad? or what are you? Have ye
> no wit, manners, nor honesty, but to gabble like
> tinkers at this time of night? Do ye make an
> alehouse of my lady's house, that ye squeak out your
> cobblers'[54] catches without any mitigation or remorse
> of voice? Is there no respect of place, persons, nor
> time in you?

SIR TOBY BELCH

> We did keep time, sir, in our catches. Be hanged![55]

MALVOLIO

> Sir Toby, I must be round with you. My lady bade me
> tell you, that, though she harbours you as her
> kinsman, she's nothing allied to your disorders. If
> you can separate yourself and your misdemeanors, you
> are welcome to the house; if not, and[56] it would please
> you to take leave of her, she is very willing to bid
> you goodbye.[57]

SIR TOBY BELCH

> "*Every time, we say goodbye, I die a little.*"[58]

MARIA

> Nay, good Sir Toby.

CLOWN

> *Every time, he arrives . . .*[59]

MALVOLIO

> Is't even so?

SIR TOBY BELCH
"I wonder why, a little."

CLOWN
Why he spoils our fun.

MALVOLIO
This is much credit to you.

SIR TOBY BELCH
Shall I bid him go . . . / and spare not?

CLOWN
O no, no, you dare not.

SIR TOBY BELCH
Out o' tune, sir: ye lie. Art any more than a
steward? Dost thou think, because thou art
virtuous, there shall be no more cakes and ale?

Figure 13.2 *Twelfth Night.* GSF 2000, Dir. Epstein: Bruce Evers (Sir Toby Belch), Chris Kayser (Sir Andrew Aguecheek). Photo: Georgia Shakespeare Staff.

SECTION IV

DURING REHEARSALS

14

DEGREE AND NATURE OF INVOLVEMENT

One of the things that should be established early is the extent of the dramaturg's presence in rehearsals and when that presence will be most clearly felt, since random dramaturgical visitations tend to minimize influence, invariably leading to situations where crucial decisions that could have used the dramaturg's input get made in her absence. The general rule, needless to say, is that the dramaturg who is most present throughout the process is likely to have the most input and be viewed as an integral member of the production team.

In fact, however, dramaturgs are often not able to commit to being in rehearsals all the time, especially if their work is on a volunteer basis. As noted earlier, the LMDA employment guidelines for freelance production dramaturgs begin from the assumption that the dramaturg will attend 50 percent of the rehearsals, though they also assume that the dramaturg will be paid 50 percent of what the director is paid, something that happens rarely in regional Shakespeare festivals. In any case, if some agreement can be reached in advance, so much the better, at least so that the dramaturg can avoid overstaying her welcome or failing to do what the company believes to be her share of the work. In setting up such an agreement, or in negotiating the issue during the process itself, some general principles should be taken into account.

I have said that the traditional work of the dramaturg tends to be most extensive and valuable early in the production's work, particularly with regard to the construction of the script, but the same idea is true of other stages in the process, including rehearsals. Indeed, the principle of early involvement applies to each subsection of the larger process of creating and staging the show. At each stage or subsectional phase of that process, the dramaturg's role changes and, in many ways, diminishes.

The rehearsal phase is a constructive arc, usually beginning with presentations, a read-through, and general discussion, moving on to other forms of table work, then to getting each scene up on its feet. When the show is blocked, albeit tentatively, focus turns to stumble-throughs, run-throughs, scene polishing, technical rehearsals, dress rehearsals, and so forth. The rehearsal phase of the production, then, is itself composed of subsections and in these too, the dramaturg's most valuable involvement generally comes early, though that work tends to decrease slightly with the move to each subsection. By the end of the rehearsal phase (tech. and dress rehearsals) the dramaturg has become a largely silent presence, a representative of the audience, an occasional sounding board, even a cheerleader. In short, the dramaturg who does not intend to be present at every single rehearsal should budget his time so that he can be around during the *beginning* of each section or subsection of the process, and should assume that those sections occurring early in the rehearsal phase will demand more of his time and energy.

As the above description suggests, the dramaturg's role changes significantly in the course of the rehearsal process, and some of that change is about becoming less visible, less audible. This is true because the conscious focus on matters of research, undergirding ideas, on theoretical concerns and all things literary (including the script itself) tends to recede once the show is getting up on its feet, becoming more about bodies and voices and light and sound. As the opening of the show approaches, and the conceptual gives way to the material, the dramaturg seems to fade, as if his contribution is largely made, the fate of the show turned over to the actors and production crew. This does not mean that he is no longer involved, but it does say that the nature of that involvement has become quieter, more mediated, the nature of the dramaturgical collaboration becoming less directorial and more audience-oriented. Dramaturgs who fail to recognize this and who, for example, give notes suggesting an entirely new way of thinking of a scene or character only days before opening is likely to produce panic and resentment. In the chapters of this section, I address different aspects of the dramaturg's role during the rehearsal process, moving in roughly chronological order through the subsectional phases as productions normally encounter them.

15

Tools of the Trade and Research Packets

Before rehearsals begin, the dramaturg's role tends to be solitary and reflective with no particular constraints on his time other than those he sets for himself, providing he meets basic deadlines (completion of a script pass or program notes). Once rehearsals begin, these luxuries vanish, and the dramaturg will be expected to produce ideas, solutions, or answers virtually on the spot. Certain resources thus become essential.

I said at the beginning of the chapter on thinking about script that the dramaturg should have several good editions to hand to serve as sources of information and to point up differences in matters of text and its readings. These editions should make the journey with the dramaturg into the rehearsal room and should be kept to hand. Rival footnotes can help to unpack a difficult or contentious passage, and differences in the texts themselves might introduce options the script could not contain but which, once the actors begin working with it, prove valuable. Editions that contain extra material (on the play's stage history, for example, or on Renaissance documents that connect to the play) can be especially useful. Actors are often remarkably good at finding points of interest and value in even the most minor of contextual details. Even when their own sense of a character or moment is fundamentally at odds with the information presented (an eighteenth-century engraving of their character, for example), having access to such information can be extremely helpful, if only because it helps the actor to clarify what he or she does *not* want to be.

Some of this extraneous material (be it historical, pictorial, technical, or whatever) might be so telling that the dramaturg will want to share it with the actors in the form of a handout, perhaps even a folder or scrap book of related documents tailored either to the group as a whole or in smaller batches for individual actors. Of course, such packets should be compiled with an eye for presenting the useful, and in manageable doses, and should be approved by the director in advance. Scholarly material should be handled judiciously, boiled down into representative quotations, for example, or presented in a short piece by the dramaturg himself (perhaps a version of his program notes, which will probably have already been written at this stage).

Actors are busy and have enough to worry about learning their lines without being buried beneath bundles of densely academic prose. Visual images are particularly helpful, because they can be examined relatively quickly, and because they evoke something palpable and concrete, something most actors find easier to connect with because their craft is itself rooted in the palpable and the concrete.

Information that actors generally find useful and that could be compiled for distribution at the start of rehearsals is generally that which can help them to conceive of their character or environment in concrete terms, so it must be shaped by the nature of the individual production. Maps of Renaissance Windsor might be interesting for a production of *The Merry Wives*, but will be rather less helpful if the production is being set in the 1950s. King James's writings on monarchy are invaluable to a literary and historical analysis of *Macbeth*, but the setting and mood of the production might render such details marginal at best. The dramaturg is constantly seeking to aid the production, not his own pet interests, and research should thus be constructed so as to be as directly useful as possible. If it is not, perhaps because it is too abstract or rooted in a historicist sense of the play, which is not relevant to the production, then the research is time wasted, the dramaturg has missed an opportunity to shape the final show and, perhaps worst of all, he has further announced his own marginality, a position from which it gets steadily harder to influence the production.

In the compilation of research material that is to be shared with the company, then, the dramaturg should answer some simple questions:

1. What are the actors most likely to find baffling or alien about the play (after it has been cut)? What kinds of research might shed some light on these problems?
2. What issues within the play text or the overarching concept—issues that are not necessarily confusing of themselves—might be highlighted, drawn into the foreground, or made immediate, by research and by research of what kind?
3. Is the research genuinely in the service of the production, or does it really serve a separate agenda that will render it either counterproductive or irrelevant? Is this particular production genuinely likely to be made better by this research?
4. Is the research marshaled and presented in such a way that those who read it will find it interesting, clear, and of material use value? Is it worded for a general audience or is problematically "professional" in its discourse, or otherwise arcane to the point of impenetrability? Is it vivid and striking? Can it be made so?

There are two kinds of useful research that the dramaturg might hand out. One is general, connected to background issues, ideas, debates, and so forth, and informs the general ambience of the show's world though it may not appear conspicuously (or at all) in the production that the audience sees.

This kind of research gives the company a handle on what they are doing, and helps them to feel that their approach to the play is anchored and legitimate, though it may not surface overtly in performance. Once, during a production of *The Comedy Of Errors*, I brought in information on Ephesus (where the play is set) as a center of early Christian struggles (theological and economic) with the cult of Artemis, this in spite of the fact that the production was to be a stylized "Pop Art" production based on a version of the 1960s. The director, Sabin Epstein, had asked me to bring in such material and I was, initially, skeptical of the value of information so at odds with the production that was to come. In the course of our initial conversations with the actors, however, the material proved invaluable because it established Ephesus's long-term association with both the mercantile and the mystical, factors that loom large in Shakespeare's play and that would make the cross-over to our upcoming show intact. Trade is everywhere in *Errors*, as is magic—or the fear of it. The production made no direct reference to the cult of Artemis, or to Saint Paul, or the origins of the Christian cult of the Virgin Mary, but discussion of such matters helped the actors to embrace an approach to the play that was unconventional in its abstraction and stylized movement, because core ideas and issues of the production were anchored in both Shakespeare's original and the history on which the play's sense of place was based.

The second form of useful research is that which has—or can potentially have—material effects on the final performance. These might be large conceptual ideas worked out with the director, or they might be details: plans of

Figure 15.1 *Comedy of Errors.* GSF 1999, Dir. Epstein: Jonathan Davis (Dromio), Peter Ganim (Antipholus), Chris Kayser (Dromio), and Linda Stephens (Balthasar). Photo: Georgia Shakespeare Staff.

the Roman Senate and descriptions of procedures therein for a production of *Coriolanus*, a careful breakdown of the social orders represented in *Henry VI Part II* and who fits where (actors often need a clear grasp on the relatively unfamiliar issues of social hierarchy in a play), descriptions of how to prime and fire a medieval canon, how monks spent their days, how to eat without a fork, and so on. All of these might be materially present in the final show in elements of what the actors actually do on stage.

The above examples are largely historical and tend to evoke either the Renaissance or those earlier time periods that are the framing worlds of some of the plays as written, but the dramaturg may well have to venture into less familiar territory in the service of the production. Basically realist shows set in the Arabian desert, or the Old West, in nineteenth-century Prussia and twentieth-century Berlin all need to be thoroughly grounded in the worlds they attempt to invoke if a sense of pastiche is to be avoided. Some of this research is likely to be done by the set and design teams, but the actors may well want more information, particularly of a kind that goes beyond the appearance of things and clothes, and they will turn to the dramaturg for such information. As with my *Errors* example above, the dramaturg can also help tie the visual design concept to the play text through research, thus aiding the sense of an organic link between script and concept. Without such a link the company can often feel that the concept has been merely "painted on," that it does not clearly "speak to" the play itself, and if this is not addressed the production can suffer, either because some of the actors begin working against the concept, or because they simply commit to it less. Here, as elsewhere, the dramaturg becomes one of the authenticating voices of the show, the person who is there to convince everyone that what they are doing is valid as an approach to the play. The dramaturg is, as I have said, an authority, but his function is not to police the production in terms of correctness so much as to use that authority to bolster the show's confidence in its own validity or authenticity. Research that makes the link between script and conceptual frame helps to make this sense of authenticity concrete, thus invisibly undergirding the show's sense of itself, and asserting a logic and coherence rooted in the performance rather than the text or author: that Shakespeare the man or his textual traces have nothing to do with the First World War in no way undermines the relevance of such a context to a staging of, say, *Henry V*. How convincingly a concept is executed and how totally the company can embrace that concept can come down to the details of dramaturgical research.

For a recent production of *Macbeth* (2004, dir. Drew Fracher), which was to have a deliberately supernatural bent and was set in ancient Scotland, I prepared a binder of about sixty pages of material to be shared with the actors, director, and design teams. None of this drew on prior productions or on literary criticism of the text except for a couple of pages of ideas listed as bullet points, targeting specific issues as they have been raised in recent criticism in concrete terms (the play's length, dating, politics, religion, ideas about equivocation, male violence, and the play's women). The rest of the source book compiled information and images—much of it culled straight from the Internet—which were tailored specifically to the production's conceptual frame. The contents page looked like this:

> Map of Medieval Scotland
>
> Bullet-point notes on the play as text in history
>
> Supernatural Scotland:
> >   Ghost stories
> >   Omens and Portents
> >   Dark Forces
> >   Mysterious Places in Scotland
>
> Scottish Witchcraft: Folklore and legend
>
> The Survey of Scottish Witchcraft: a sociological history
>
> Life in a Medieval Castle
>
> Program Notes
>
> Images
>
> Scottish Landscape
>
> Scottish Castles

In each case, the information was geared to the play (the map pinpointed every place name mentioned in the text) and, more specifically, to the production.[1] The section on Supernatural Scotland combined myths, folk tales, and ghost stories from various periods and places in Scotland, not so much to feed the production directly as to instill in the company a sense of the show's world and cultural climate, one that rendered the play's mystical elements more real. Likewise the images were carefully selected to reinforce a sense of that same world's rugged or ethereal strangeness. Some of the material produced stage business directly (the Life in a Medieval Castle section had notes on table manners, hygiene, and social decorum, which came to inform the banquet scene directly), but much of it produced less palpable—but no less valuable—results, aiding the actors in investing mentally and emotionally in the staged environment and its attendant beliefs and practices. Much of this material has little to do with the way the play would originally have been read, less to do with how it would have originally been staged. Nothing in the play convinces me that Shakespeare had ever been to Scotland or knew anything at all about medieval castle life. That doesn't matter. The research is aimed at grounding the present production and those involved in making it through contextual detail and usable (however indirectly) specifics, and its value thus has little to do with Shakespeare the man, his period, or even, strictly speaking, his text.

Once in rehearsals, new questions or points of debate will likely arise and these might require the dramaturg to do further research outside the

Figure 15.2 *Macbeth*. GSF 2004, Dir. Fracher: Sherman Fracher (witch), Marni Penning (Lady Macbeth), Daniel May (Macbeth), Alison Hastings (witch), and Bruce Evers (witch). Photo: Bill DeLoach.

rehearsal hall. More often, however, issues can be addressed according to what has already been compiled and through reference to other source materials that the dramaturg has to hand. As well as the textual editions already mentioned, the dramaturg should have the following:

1. A good complete works of Shakespeare, for access to other plays (Bevington, Riverside, Arden, Norton, and the like).
2. A glossary of Shakespearean words and phrases such as that by C.T. Onions (first published 1911, revised in 1919, and subsequently reprinted) or the more recent volume by David Crystal, Ben Crystal, and Stanley Wells.
3. A good (extensive) dictionary: preferably an *Oxford English Dictionary*, ideally the complete, multivolume version (edition 2 or 3). Many academic institutions have on-line versions of the *OED* to which the dramaturg can get access through a laptop computer (quietly) during rehearsal or from a terminal close by. The *OED* is unrivalled for its depth, its etymological detail, and its explication of words that have fallen out of use.
4. A copy of the edited play script with plenty of space for annotation.

It is also important for the dramaturg to get hold of anything other people in the cast are reading that may shape their participation in the production, particularly if there is reason to doubt its accuracy or agenda. Student actors, for example, sometimes turn to high school study guides or websites to get a handle on the play, and these things tend to be at best, inconsistent and, even where they are not actually wrong, subjective and simplistic. They tend, moreover, to suggest a single way of viewing things in the play, something that can become a nuisance if the production is moving in a different direction.

More mature actors are likely to have their own favorites when it comes to research, and while some of these are extremely good, they tend to feel dated from an academic standpoint.[2] Even some of the recent publications that tend to dominate the nonacademic critical perspective on Shakespeare (such as the work of Harold Bloom, for example) are often significantly out of step with current trends in critical theory and criticism. The dramaturg who comes from a literary background may find this off-putting, but she should feel free to voice such disagreement after she has, of course, looked over the contentious material herself. Generally, I think, the more actors read about their parts and the play, the better, regardless of what it is they are reading. Such reading only becomes problematic if it leads the actor to hold on to ideas that are palpably at odds with the show as it is being conceived, or with the work of their fellow actors. In such a situation, it might be appropriate to provide (either in print or simply through conversation) alternate—even contrary—perspectives.

Knowing Your Audience: Talking to Directors, Talking to Actors

It is an unavoidable component of the dramaturg's role that she is always, as Cary Mazer has said, "in the background [. . .] silent, pensive and slightly out of focus" ("Solanio's Coin" 10). If text and performance are fundamentally—even generically—different, it is the dramaturg who embodies that difference, the person who has to see her job as having roots in two quite different camps, one textual, historical, and intellectual, the other performative, immediate, and kinetic. As I said at the end of part 1, these two camps speak significantly different languages, and one of the results of the dramaturg's strangely bifurcated role is that she has to be a competent speaker of both. As with any linguistic proficiency, one language is likely to be somehow prior—absorbed through one's experience—while the other is secondary—studied. Given the fact that many dramaturgs are literary critics first and come to practical theatre thereafter, it is more often the case that dramaturgs speak the language of theory, history, and text as a "first" language, and so have to work at translating their ideas into the idiom of the rehearsal room.

There is a danger here of oversimplifying, suggesting that all actors and directors conceive of their work in the same way and thus use a uniform vocabulary, which is far from true. They vary according to their training, experience, and personality, some being of a markedly academic bent (particularly those attached to educational institutions), some being highly political in their sense of the plays, some having a sense of theatre that is rigorously theorized or pointedly postmodern. Others—much more commonly, I think—see the plays as being a largely naturalist environments in which characters they assume to be well-rounded—even possessing personal histories that precede the portion of their lives detailed by the text—pursue their conscious and subconscious desires. The actors and director seek to fully understand, even to internalize, these desires during the rehearsal process. Whatever one thinks of the virtues of these various methods (or Methods) or their suitability to Renaissance drama, the dramaturg thus needs to adjust her vocabulary in order to be of use. In print (in the aforementioned research packets, for example) and in conversation, the dramaturg must recognize her audience and speak accordingly in order to stay comprehensible and relevant. This is not about "dumbing down." It's about translation. The dramaturg who wants to have a real impact on a show needs to embrace that distinction absolutely.

Whatever the dramaturg's own background and training, his role is defined largely by the use of an intellectual faculty, and as such he tends to be comfortable with a more abstract or esoteric mode of discourse than most actors and, to a lesser extent, directors. Again, this is not a matter of intellectual capacity so much as it is about perspective and focus. Actors are generally trained to see the play through their character, and their engagement is thus more kinetic, emotional, and tactile. The dramaturg (particularly one with literary critical training) sees the play if not as a network of ideas (which is often—and usefully—the case) then at least as a larger entity, a structure or structures to which the actor immersed in his or her character is generally too close to see. In broadly realist productions, the actors have to be saturated by the world of the play not in theoretical or ideological terms, but in the material dimension of such theory and ideology, the concrete conditions with which the characters engage daily.

For example, imagine the rehearsal process for a dark and highly politicized production of *Measure for Measure*. The dramaturg has been talking to the director for some time about a way of viewing the Duke's disguised monitoring of his people in the expressly Foucaultian terms of surveillance and Panopticon, coupled perhaps with a sense of the emerging Renaissance discourse of conscience and the political/theological rhetoric of an all-seeing deity.[1] The world of the play is one in which an autocratic state government polices the very thoughts of its people. If the dramaturg is going to talk successfully to actors about this concept during rehearsals, he would probably do well to emphasize not the larger ideological frame (the theory of how a culture internalizes its sense of right and wrong for reasons as much bound to issue of social control as about morality) but the *manifestation* of such an ideological frame in its citizenry. The dramaturg might ask the actors to imagine living in Stalinist Russia, or facing Macarthy's inquiries into un-American behavior. He might suggest that they consider some aspect of their private selves to be forbidden and punishable if it were discovered, or that there may be bugging devices concealed even in public places. Part of the dramaturg's job is to get the actors to understand and invest in the larger concept then, and he does so by giving it to them in terms they can use and in ways that open up possibilities rather than closing others off. I like to use questions: "Was your character followed on his way here? Are you sure? Can this person you are talking to be trusted? If the government offered him money to inform on you, could he still be trusted? What if they offered you the money to inform on him?" Some actors thrive on an intellectual and theoretical investigation of the play and the show, but many don't, and they don't need it to make such intellectual and theoretical concepts playable and recognizable to the audience.

These examples raise an important question: when does the dramaturg start behaving like a director? The answer depends on the show and the company, of course, as does the sense that this is necessarily a bad thing. Some directors welcome direct input from everyone in the room, and by "direct" I mean unmediated, passed directly to the actor as the dramaturg (or whoever) thinks of things to say. Other directors prefer that only they talk directly to the actors, everything else going through them, and that sparingly. While the first situation seems ideal, it can bog down the process, and I would advise the dramaturg to make use of such freedom carefully, with a sense of the director's goodwill, which could be withdrawn if the dramaturg seems to be subverting the director's authority or ideas, or if things are moving too slowly. As I said earlier, the dramaturg's role as a constructor diminishes as the process moves forward and there is usually a time (often shortly after every scene has been blocked) that further discussion and extensive redirection becomes unwelcome.

The second alternative, in which the director mediates all comments from the dramaturg (and of course many directors, even most, prefer some kind of position in-between these two extremes), seems controlling and repressive, a dynamic that shuts the dramaturg out of the process, but this is not necessarily the case. Indeed, sometimes this kind of mediation can be a dramaturg's best friend, and for several reasons. First, it makes it less likely that the dramaturg will be seen to make suggestions that are—either in matters of tone or content—inappropriate or impractical, thus saving him some face. Second, his suggestions get the automatic authority of the director and are more likely to be taken seriously. Let me say again that the dramaturg's allegiance is to the show, not to his own ego. If a good idea can be brought to bear on the production only by going through the director then that is the route the dramaturg should take. Whether or not the director credits the dramaturg with the idea as she presents it to the actors is of no consequence, and a skillful dramaturg can sometimes get a director to think that she came up with the idea herself! More often than not, the director need not be dodged or manipulated like this, and generally should not be, but there are real issues of rehearsal room protocol here, and the very worst thing for the dramaturg's cause is to be seen as somehow undermining the director or as trying to redirect what the director has done. The dramaturg thus strives to ensure that his ideas get a full hearing, perhaps get tried out in rehearsal at least in discussion, but if the director finally drops them, nine times out of ten they should stay dropped. The dramaturg should never leave the actors feeling that they are being given contradictory advice, or do anything to suggest that the director and dramaturg are not finally unified on the most important issues of the show.

Dramaturgs unused to the rehearsal room environment often make simple mistakes, which, if not actually giving offense, can jeopardize their own credibility. The following are common examples:

1. The dramaturg spends all his time reading and writing. Rehearsals are about voice and action, so the dramaturg needs to watch and listen. Failure to do so suggests that he assumes everything of value exists on paper and not on stage.

2. The dramaturg sits in the wrong place. Rehearsal room conditions vary, of course, but the dramaturg should not be seen to be physically usurping the director's position, nor should he be self-marginalizing in ways that take him out of the proceedings (by sitting in a corner, for example, where he can't see properly or talk to anyone). Early in the process, the dramaturg should stay close to the director whenever possible but will generally remain seated when the director starts moving around. This way he should have the opportunity to make suggestions without getting in the way. If in doubt about where to set up, ask the director.

3. The dramaturg broadcasts her separateness from the expressly theatrical work being done. It can take time to feel comfortable in the rehearsal room if one is unused to it, but the dramaturg should try to appear in her element, something easily undermined by a failure to follow the conventions of the locale. Actors and directors, for example, often dress very casually for rehearsal. The dramaturg who comes dressed to lecture immediately suggests, at best, marginality, at worst, superiority: another incarnation of the Shakespeare police.

4. The dramaturg offers line-readings or tells actors which words to emphasize ("Do it like *this . . .*"). Actors want options, not answers, and to make suggestions premised on a notion of correctness invariably alienates the cast who feel that they need to be able to make interpretive choices in order to fully invest in the scene.

5. The dramaturg offers irrelevant information or research and at times when the actors are focused on other things.

6. The dramaturg is dialectical to the point of total ambiguity. Actors want options, but they also want specific responses. If the dramaturg's suggestions are wholly baffling or full of second-guessing and qualification (often required in literary academia), whatever is being offered tends to recede into grayness.

7. The dramaturg is self-effacing to the point of invisibility. The rehearsal room is a place of performance. While there is value in self-deprecation, the dramaturg is present to function as an expert, something seriously undermined by an overabundance of humility.

8. The dramaturg is condescending, wordy, abstract, rude, distant, star-struck, or otherwise amateurish.

9. The dramaturg promotes faction, either between the actors or involving the director.

10. The dramaturg is never there when he is needed or comes unprepared. If the dramaturg cannot answer a question based on his current knowledge and sources, he should address the problem with some research and be ready to offer an answer at the next rehearsal or by e-mail or phone beforehand.

11. The dramaturg talks while work is being done (and whether it's chitchat or earnest textual discussion doesn't much matter to the actors who can't concentrate).

Above all, the dramaturg should manifest a genuinely collaborative spirit, one that makes the actors comfortable and inventive, not leaving them to anxiously (or belligerently) await arrest by the dramaturgical Shakespeare police.

Of course, even the right attitude can be misread and the dramaturg should thus establish as quickly as possible a clear understanding of the protocols and dynamics of the rehearsal hall. Stage managers (who also exist on the peripheries of things, if in different ways) are usually masters of these codes and the dramaturg can learn a great deal by watching how they work, particularly in the way that they raise concerns or questions of the director during the flow of rehearsal. If in doubt, the dramaturg should ask the director and should, periodically, give her the opportunity to pass comment on the dramaturg's work and demeanor to date, at least until some kind of smooth working relationship is established.

Where talking to directors is concerned, the dramaturg generally has more freedom than in talking to actors, because the director, like the dramaturg, sees the entire play, not just the parts of it a single character sees. As such, the director is likely to be more interested in how things might play to an audience than are actors who are often more concerned with their characters' internal motivations, and this is an excellent way to discuss larger aesthetic or political concerns. Such large debates work best outside rehearsal room conditions, however, where things tend to be detail oriented, and I find regular e-mail bulletins to the director to be especially useful as a forum for discussing possibilities or what is and is not working. Cary Mazer's description of his e-mails to the director of *The Merchant of Venice* production for which he worked as guest dramaturg are especially illuminating for their frankness, convivial spirit, and intellectual stick-to-itness ("Solanio's Coin"). E-mail is less intrusive than phone calls and encourages a clear expression of ideas in print, while maintaining a sufficiently informal mode to make them conversational. So long as the director is checking e-mail regularly, it is an excellent way to maintain a sense of debate and involvement from the dramaturg without impinging on rehearsal time.

17

The Dramaturg in Rehearsal: A Temporal Breakdown

As I have already said, the dramaturg's role changes in the course of the rehearsal period. This section briefly maps the nature of that change, on the understanding, of course, that different companies behave according to slightly different models. The rehearsal process performs a steady shift of emphasis enacting a series of crucial transitions from the theoretical to the material, and from the potential to the concrete. As opening approaches, of course, productions become more rooted, less spontaneous, more repetitive, less improvisational, and while they strive to keep some of the earlier less predictable energies of the show, they move closer to that sense of being finished, polished, and ready to be seen with some degree of consistency. This does not mean that thought and creativity wholly give way to mechanics and memorization, but it does require that the dramaturg cease bringing up ideas or possibilities as she might have done at the beginning of the process.

In general, the dramaturg's role in rehearsals shifts, as does the production's focus, mirroring the overall move from exploration and discovery at the start of the process to presentation and fluency at the end. In short, the dramaturg changes from being one of the production's originators (in which her function is attached to the director) to being one of its monitors (in which her function is as a stand-in for the audience), a change that requires a steady move from the backstage conditions of the rehearsal hall to the front of house conditions of tech. and dress rehearsals. Simplistic as it may sound, much of this change can be mapped according to where the dramaturg sits.

Introductory Remarks and Read-Through

Most productions begin the rehearsal process with a day or two that include presentations by the director and designers, proceeding to a reading of the script by the cast, and some initial discussion of the project. This important and exciting stage allows everyone involved to get acclimatized to the conceptual frame of the production and to each other, and the dramaturg is often called upon to help make this happen by presenting some introductory remarks. Obviously, the nature of these remarks will vary tremendously according to the nature of the show, but some general pointers can be made that will apply to most cases.

In this first meeting or meetings, the cast and representatives of the design team and production staff may be assembled together for the only time until technical rehearsals begin, so everyone has an opportunity—and a responsibility—to make remarks that can include everyone present. Time tends to be precious in these first meetings, so the dramaturg should go in prepared to utilize her slot (say five to twenty minutes) as effectively as possible, and this will require some preparation. Rambling, loosely arranged thoughts may seem to work adequately in the moment, but a dramaturg who misses the opportunity of addressing this captive audience in a more organized fashion will regret it later, when she finds herself repeating the same anecdotes and ideas to individual members of the cast because she didn't tell the whole group when she had the chance. This first presentation is, furthermore, a crucial moment for the dramaturg, particularly if she is not well known to the company from previous work. For many of those in the room, the dramaturg may be an unknown quantity, one whose role and perspective is unclear, a role that ("Everybody stop what they're doing . . . . The Shakespeare police have arrived") can produce nervousness, skepticism, and defensiveness. The dramaturg's introductory statement can go a long way toward allaying such negative feelings, thus preparing the way for a more profitably collaborative experience. He or she should be clearly a member of the team and engaged in all debate that is generated in these first meetings. This means sitting at the table with the director and actors: an active and invested collaborator, not a bashful academic in the corner.

The dramaturg's remarks should be generally cleared with the director in advance, if only to save time and avoid repetition. Contradiction should not be an issue at this stage, because the dramaturg and director are a team working on a shared vision. While a respectfully mutual antagonism is sometimes useful between dramaturg and director, they need to present a united front to the cast and crew at this stage. Since the director will be doing most of the talking, and either giving the designers a forum for them to present their ideas or serving as a mouthpiece for these ideas, the dramaturg's remarks should address an issue or two that forms a subset of the production's conceptual frame. Since the dramaturg's territory is the intellectual engagement of text, context, history, and theory, these areas should feature prominently in these first remarks in ways that are clearly relevant to the upcoming show.

A good place to start is with some brief observations about the script as it has been prepared, outlining any conspicuous changes or deletions, presenting the rationale for such modifications, and asserting the integrity of the script as a starting point for the work that will follow. It is important that the actors have faith in the edited script and do not spend the first weeks of rehearsal second-guessing its choices, so the dramaturg needs to underline its essential authority, something that may be helped by the *limited* application of some general theoretical premises. As ever, the dramaturg should show herself to be an authority without seeming to be either a disconnected intellectual presence, or someone who is too clever or intimidating to talk to.

A fruitful second phase is to proceed to a core idea or issue, something integral to the intellectual aspect of the play and the production, something that might be nicely fleshed out by a short handout or an image or two that can be circulated as the dramaturg speaks. Conversely, this might be an opportunity to draw specific attention to materials handed out as part of a general research/background packet. For example, a dramaturg's introductory comments for a production of *The Taming of the Shrew* might present a quick overview of Renaissance discourse on gender relations and hierarchies as manifested by contemporary ballads, homilies on obedience, radical pamphlets, or tracts on related subjects such as falconry.[1] This is complex material, of course, and I am not suggesting that the dramaturg spend an inordinate amount of time on it, rather that he or she present a general sense of the idea. This might be as simple as an articulation of the period's plurality of perspectives on the issue, or an argument about the way notions of gender were tied to—even subsumed within—other discourses dealing with theology or state politics. The referenced documents can then be left to be considered by the cast thereafter.

Likewise the dramaturg's introduction to a production of *Macbeth* might fasten on Renaissance ideas about kingship if they are appropriate to the production's approach (perhaps including some pithy lines from James I's *Basilikon Doran*).[2] Conversely, the dramaturg might discuss the kinds of materials I already mentioned as being part of my research for a recent production of *Macbeth*: Scottish landscape, folklore, castle life, and so on. The discrepancy between these two ways of introducing the play to the cast manifests two general approaches to such introductory remarks, one that is historicist and seeks to provide information based on how the play might have been originally perceived, another that is more interested in creating a world or ambience for the production based on what the play *can* mean. As I said before, there is little in the play as written to suggest that *Macbeth* is crucially built upon a real sense of medieval warfare and culture, but a production that is set in or around the time that the historical Macbeth lived might make real use of such things and gain a specificity that is no less valuable for being of little relevance to the English Renaissance.

Different from such a case by degree rather than kind is a production such as that of the *Julius Caesar* for which I served as dramaturg (dir. Dillon), which was set in 1930s Louisiana. Much of my introductory material for this show concerned Hewey Long, and I dipped into the Renaissance only to set up and justify by analogy the use of this seemingly anachronistic setting by arguing that Shakespeare had done something similar in making a tale of ancient Rome ultimately "about" contemporary London. This does not mean, of course, that our story would be the same as that told on Shakespeare's original stage, just that we, like him, were drawing on history (in our case, four different histories linked to Rome, London, Baton Rouge, and Atlanta) to make a story fresh. We thus sought points of proximity (similarities, not replications) between the various periods under consideration as a way of making the story innovative and compelling. In this instance, conventional historical scholarship about the Elizabethan discourse surrounding the story of Caesar nicely bolstered—and lent authority to—the production's less conventional setting. The goal, as I was at pains to explain, was to take what was powerful and valuable from Shakespeare—particularly from the words of the dialogue itself—and reframe it, using the other historical and geographical referents to throw the story into new light in ways that would trigger new connection and revelation in a contemporary audience.[3]

Because the dramaturg serves neither history nor the original text but the production itself, he or she is an excellent person to bring research and intellectual energy to the explanation and justification of the production's framing concept. Working on a production of *The Winter's Tale*, therefore, I was able to discuss in conventionally academic terms the play's use of the Bohemian seacoast and the Renaissance's sense of the Delphic oracle to help construct our production's world, despite the fact that that world was grounded neither in ancient Greece nor the English Renaissance.[4] The specifics of these cultures fed more general ideas about the Providential universe of the production, its aura of metadiscourse, and a nonnaturalistic theatricality.

Sometimes it is helpful to map key elements of plot or backstory at this early stage of rehearsals. Chronologies and family trees are especially useful in the histories, for example, even if the production is finally not especially interested in telling a quintessentially English story, and they make excellent first

Figure 17.1 *Julius Caesar*. GSF 2001, Dir. Dillon: Theresa DeBerry (Calpurnia), Brik Berkes (Decius Brutus), Bruce Evers (Julius Caesar). Photo: Rob Dillard.

day handouts. Such things are useful for clarifying convolutions of plot, indicating the period that the play spans, or explaining crucial facts that ghost the story and which the original audience would have known (the significance, for example, of the usurpation of Richard II to the later plays of the Henriad). It is always gratifying to hear actors say after the show has opened that some dramaturgical point made in the first week of rehearsal switched a light on in their heads or somehow rendered clear what had been hitherto baffling. The dramaturg's part in these early meetings can sometimes trigger a communal sense of discovery, which will resurface periodically throughout the process, or help to anchor it. In a production of *Cymbeline*, for example, a chance remark about the play's "down the rabbit hole" version of truth, which had found its way into a draft of my program notes, became part of the production's guiding lexicon, popping up from time to time throughout the rehearsal period as the actors steeped themselves in the show's surreal logic and imagery.

The read-through of the script requires little of the dramaturg except that he listen. This is probably the first time he has heard the words upon which he has worked spoken aloud and it is often extremely provocative, not least in suggesting how clearly the plot, characters, relationships, issues, themes, and ideas leap off the page. The dramaturg should annotate his script as the reading progresses, marking those passages or moments that seem now to present difficulties. This will also be the first opportunity to see how the actors function, both as a body and as individuals, and this may suggest the need either for more work on the script or for conversation with the director about some of the actors' initial approaches. Of course, many actors will go on an extensive journey from the first read-through to opening night and their character will shift massively, but with some (and here familiarity with the company and the insight of the director are especially helpful) the initial read is a good model for what the actor will finally bring to the role. As such, things that clearly don't work—or don't work as well as they might—in the first read should be given real attention immediately.

The read can also provide a reliable sense of how well the framing concept will work. Sometimes, in initial discussions, certain points can loom larger than they actually seem to be when the script is read, and sometimes the director and dramaturg can misinterpret aspects of the text, so that the production gets balanced upon shaky ground. Identifying such problems now—and remedying them—will make for a considerably smoother and more self-assured show later.

This is also a good time to think more precisely about running time (the stage manager will probably clock the run), since any major cuts or changes to the script should be made before the actors start scene work. Likewise, if it hasn't already been decided, this is a suitable point to talk to the director about when to place the intermission, a minor decision but one that can have real impact on matters of story structure, dramatic pacing, and climax.

After the reading, the dramaturg should make a point of meeting with as many of those involved as possible, to introduce himself personally and to ask what they think of the script. Actors' observations (what they found confusing, changes they think are overly modern, edited lines they miss, and so forth) can help the dramaturg to see things that need his attention or that should be put on the director's table. At this stage, I am particularly hopeful that actors will talk about the cleanness and clarity of the script without a sense of it being thinned or reduced unduly (and I mean that as much about the script's seriousness as I do about its length). If there's a lot of talk about the script's difficulty or opacity, another editing pass might be in order.

Throughout these presentations and conversations, the dramaturg should be engaged and enthusiastic. He should seem committed to the project (again, not just to the play) and keen to collaborate. He should be supportive and tolerant of other people's ideas, even when he doubts their usefulness or accuracy. He can explain his points and he can disagree, but he must do so with respect, with humor, with warmth, with the sense of being a fellow traveler on this particular voyage, not a captain, not a judge, not a policeman. For the actors who are being asked to trust the dramaturg from here on, this is, after all, his audition.

TABLE WORK

In the first days of rehearsal proper, the actors will generally begin their collective exploration of the play through sessions centering on close scrutiny of the text, often as they sit around a table with the director, reading and annotating their scripts. Though some directors prefer to dive headlong into the material and have the actors make their discoveries physically as they move around and interact with each other, even having them improvise the gist of a scene, a version of table work still takes place either in that moment or at some other time. Whether or not table work involves actually sitting round a table, it does involve sitting around the script, exploring it, testing it, experimenting with it, seeing what it contains or permits, seeing what energies it can help to generate, finally propelling the actors away from their books and into the world of the production.

For the first days (or, if one is lucky, weeks) of the process, then, the script is not merely a prop for the actor to consult when she forgets her lines but is to a certain extent the root of all that follows. Despite the core truth that theatre originates in more than mere text, most actors at this stage of the process look to the words of the script to generate—or at least to authorize—their work, and it therefore makes sense that their work is viewed *at this point* as more interpretive than constructive. However radically actors will want to build their roles in ways that disregard (quite rightly) a sense of somehow fulfilling a *correct* reading of the text, they will generally want to ensure that their sense of their characters, the action, the ideas, and emotions in play, are all in accord with what the script actually says. Not surprisingly, then, this is where the dramaturg comes into her own, answering questions, offering insight, floating possibilities, and being, in general, a kind of guide for the actors as they explore the play. As a guide, the dramaturg will have continued access to research, history, and theory, but much of this initial guidance will come down to the most basic level of textual analysis: determining the sense of utterances, piecing together elements of plot, clarifying that which (your best editing notwithstanding) inevitably remains obscure and so on. Some actors will want to hold the guide's hand, others will barely notice her presence, most will be glad she's there to aid the process and broaden the options.

Right after the actors first pick their way through a chosen scene, there is often a period of discussion in which the dramaturg should try to get any big ideas or observations into the mix. This is perhaps the dramaturg's most talkative spell, though it should give way to briefer comments or observations as the session proceeds, usually in response to questions, and the ideas or observations should always be anchored by the production's stylistic vocabulary.

For the actors, this is the stage at which the dramaturg will have the greatest involvement and she should consequently be direct and unapologetic. If there *is* a table around which the table work will take place, she should sit at it. Though she shouldn't clutter the table with reference materials, she should have them on hand; while discussion will probably reach into large conceptual areas, which would be well supported by real research, it will generally stay close to the words of the script, as will most of the questions the dramaturg will be called upon to answer. Having said that, the dramaturg should be prepared to talk about textual matters in expressly human and material terms, something academic dramaturgs (myself included) can find something of a shock to the system, used as they are to thinking about the plays as loci of ideas and issues. This is particularly true of actors whose training or inclination is broadly naturalist, and the dramaturg should therefore expect questions pertaining to the characters as real people in real situations: "How old am I?" "How much time has passed since the last scene?" "Are we indoors or outdoors here?" And so forth. Often, the answer to such questions is "it doesn't matter," but for most actors attempting to make the scene feel real it *does* matter, and no amount of discoursing on the non-illusionistic Renaissance stage's lack of interest in such things will help. The modern stage is a different entity entirely and it is for the modern stage that the dramaturg is working. A less peevish answer to such questions might be "the text doesn't say," since that frees up lots of options without dismissing the (legitimate) concern, though the dramaturg had better be sure that the text really doesn't say. Dramaturgs used to thinking in the abstract sometimes disregard the concrete details that actors seize on. To get such things wrong undermines the dramaturg's credibility and creates possible contradiction and difficulty later on.

Concrete concerns often matter very much even in relatively abstract discussion of the play, and the dramaturg should thus be ready to furnish information on issues such as characters' jobs and status: What is a "botcher"? Which of *A Midsummer Night's Dream*'s mechanicals would be considered the most socially respectable? How common were knights and how were knighthoods attained? This character claims to make fifty pounds a year: how much is that and how common would such an income be?[5] This exchange seems to require the removal of a hat: what is the etiquette in terms of men wearing hats in the presence of ladies/their social superiors or indoors? And so on. These may seem trivial details, but they are the stuff of which actors build a concrete sense of person, place, and time. They are part of how the show will create meaning, and should be looked to accordingly.

Actors often find issues of social hierarchy unfamiliar but playable, the historical insight into rank creating a baseline for how characters might relate to each other.[6] Similarly, historical information on social customs and rituals—particularly those surrounding the superficially timeless points of connectivity between our period and Shakespeare's—love, marriage, birth, sickness, and death, can not only give a sense of authenticity to historicist productions, they can underscore the action of a script in surprising and revelatory ways, which open up playable options in the most contemporary or eclectic of shows. Less concrete, but at least as useful, is insight into relevant belief systems, religions, assumptions, and ideologies, which might undergird entire plays or merely supply a little filigree for a character. The good dramaturg will anticipate the relevance of some of these issues, and should therefore be ready to offer thoughts on them as soon as they come into play. Others will come up in questions or discussion and may require a little outside research before an answer can be given. The dramaturg should try to get these answers quickly, before the questions get shelved and rehearsals move in a different direction.

Other questions will be about meaning, either about the nature of an event in the story or—more simply still—about sentence-level utterance. No amount of editorial polishing of the script can render every single line crystal clear to all readers, particularly since all lines are open to various readings. Apart from the "what am I talking about here?" kind of question, the dramaturg will also be expected to field the more slippery "can it also mean X?" or "what is the tone of this word? Familiar? Abusive? Affectionate?" Again, the best answers to such questions are accurate (historically, etymologically, culturally, and the like) while opening up options rather than shutting them down, and the dramaturg should always respond as if the question is perfectly reasonable. Dramaturgs always run into moments when intelligent and experienced actors own up to not knowing what a familiar word really means, and any response should be generous, and that includes the tone as well as the content of the answer. When actors reveal themselves to be vulnerable—or are given notes that render them vulnerable—the dramaturg should be gracious and sympathetic. Good dramaturgs often begin sentences with "Yes, you'd think it would mean that, but actually . . ." or "For the longest time I thought it meant that too, but then I heard that in fact . . ." The dramaturg/actor relationship can be a very fragile one, and a little tact, a little ordinariness, a little self-deprecating humor, goes a long way. No one likes a teacher who rubs your nose in how little you know, and since very few dramaturgs can begin to do what a moderately talented actor does daily, I think that any sense of superiority on the dramaturg's part is grossly misplaced.

All this said, it is easy to offer too much information, and the best dramaturgs tend—as ever—to brevity and clarity, particularly as the table work (which itself has phases) begins to progress toward preliminary blocking. Realist actors often feel the need to temporarily "forget" all the intellectual and historical material that they have received from people like the dramaturg, putting it all on an internal back-burner, so that they can focus on the characters as recognizable people. The dramaturg needs to be sensitive to this impulse: acting, when all is said and done, is a remarkably naked endeavor, particularly in the earliest stages of rehearsals when actors are experimenting. Overly intellectual contributions from the dramaturg or—worse, much worse—condescension, dismissiveness, or mockery can severely jeopardize the rehearsal and alienate everyone involved.

Another dramaturgical function that is especially important at this relatively early stage is the analysis of the scene as a structure with arcs and beats, crises, energies, and shifts. Actors and directors are usually expert at plotting such movement but the dramaturg's sense of such things is likely to be welcome—and is often slightly different in nuance. Similar structurally based observations work equally well for subsectional units of the scene, passages of dialogue, individual speeches, and even single lines. As well as matters of semantic content, the dramaturg should pay special attention to the form and aesthetics of the utterance. Noting shifts in meter or a sudden movement into rhyme can prompt an actor or director to devise an explanation for the textual crux that takes the formal issue and makes it part of the scene's dynamic in psychological terms. At this stage, even the most textually minded of observations can produce suggestive questions, which, in turn, lead to real insight rendered in an expressly theatrical language.

Of course it should be said that sometimes such observations produce nothing of value, or—more problematically—nothing that the actors or director see as valuable. In a recent production of *The Tempest* we were doing table work on Prospero's speech in which he anticipates the relinquishing of his magic, a speech in which he details all his power has let him do, including the waking of the dead. I wanted to know whose grave he had opened up. It is often said that the speech reveals a darker side to his magic than the audience has seen thus far, and since this particular production was embracing some of Prospero's darker, more controlling tendencies, I thought it interesting to wonder if the grave he had opened belonged to the only person we know to have died on the island: Caliban's witch mother, Sycorax. Here, I thought, was the great manifestation of Prospero the humanist turning over to the Dark Side, his power corrupting him, rendering him the very thing he has defined himself against, as he communed with (and learned from) his monstrous female opposite. It was, I thought, playable, insightful, brilliant.

No one else thought so. They listened politely and moved on. I dragged the issue back on to the table a couple of times, but no one went for it, and eventually I buried it with the undisturbed corpse of Sycorax. Part of me still likes the idea, though I can see that it probably wasn't playable, not without interpolating a scene in which we saw Prospero swapping spells with a bedraggled mummy, a problem everyone else grasped intuitively. Even if it was playable, it was off topic, a direction the show wasn't going in, a schematic and overly flashy comment on the action, perhaps, rather than something that clearly belonged in this particular version of the story. In any case, it didn't fly, and I had to abandon it.

Everyone in theatre does this from time to time—actors, directors, and designers included—but it sometimes seems that no one has to jettison more good ideas than the dramaturg! This is partly the paranoia resultant from having a less conspicuous stamp on the show (and thus wanting to be able to say "that bit there was mine"), and partly the fact of not being in a position to simply make an idea happen. Directors can force the issue, actors can do their thing till everyone accepts it. Composers can say "let me play it to you and see what you think." Costume designers can say, "I went with this fabric because it seemed right. I can change it if you really want me to but it will be expensive . . . ." Dramaturgs have no such power to make ideas real. They can only offer, or perhaps lobby, but they can't *show* how such an idea would play without the complicity of others. The best the dramaturg can hope for in such a situation is that the director says, "OK, we'll try it." Explanation and enthusiasm will carry an idea only so far. The supreme test is whether it will (or might) "work" on stage, and that is a very subjective matter indeed.

The only consolation (other than dolefully singing The Rolling Stones' ever pertinent "You can't always get what you want") is that ideas which feel pasted on to productions are invariably painful to behold. If everyone is not behind an approach, if the actors and the rest of the production staff don't fully commit to that idea, it should probably be abandoned, if only because its execution is likely to be half-hearted. Since no production can be definitive, each only able to air one set of choices, it is inevitable that many ideas will go unused. Future productions will, perhaps, revisit these ideas; maybe the next *Tempest* I work on will build the tomb of Sycorax center stage.

The dramaturg's status in any stage of the process is often indicated by how central are the tools of his trade. When everyone is huddled over a table covered in scripts and reference works, the dramaturg is a central, even a leading, player and will probably be able to speak up whenever he has something pertinent to say. When the table and its piles of texts gets moved to the side in order to free up the rehearsal space for movement, the dramaturg goes with it. Even at this early stage, it is important that the dramaturg lower his profile slightly the moment the actors start picking their way through the scene physically. The dramaturg shouldn't interrupt, and should contribute thoughts or suggestions only when there is a natural break in the action, when the actors are already out of character, something that will happen a lot at this stage. The dramaturg is still very much involved as a valuable collaborator here, but he needs to pick his moments more tactfully from this point on.

Throughout the table work the dramaturg needs a certain communicative tact. Being a textual expert can be a double-edged sword since actors sometimes rely on the dramaturg for answers instead of exercising their own creatively interpretive skills, turning to the dramaturg for solutions rather than exploring the text themselves. This is one of the reasons that directors are sometimes wary of even good and collaborative dramaturgs: they continually seem to embody the literary, authorial, or historical authority that trumps the theatrical process and offers the illusion of correctness.

From time to time, then, it is appropriate for the dramaturg to be less than forthcoming even in dealing with actors' questions about what words in the script mean, or, conversely, to answer such questions with as many possibilities or other questions as possible, thereby emphasizing ambiguity. Sometimes the best course of action is *not* to be seen as the fountainhead of truth, lest the actors suspend their own discovery process in favor of bowing to ideas they assume are better because they are touched by the dramaturg's authority as a textual specialist. Naturally the dramaturg does not want to do this so frequently that her own knowledge is brought into serious question by those who want to trust her judgment, but the exploratory conditions of the rehearsal room sometimes suggest that the best thing for a dramaturg to do in response to an actor's query is to put the ball back into the actor's court. This way the dramaturg continues to divest herself of the idea that her authority is bound not to her expressly theatrical work, but to those ultimately antitheatrical modes of authority derived from Shakespeare's iconic and textual traces in the culture as a whole. As well as reminding the actors of their own self-authorizing work then, the refusal to offer answers or simple translations, further deconstructs the idea of the dramaturg as policeman, in turn promoting a more genuinely collaborative and innovative rehearsal energy.

ON ITS FEET

The table has gone, or has been shunted to the side, but the books, particularly the scripts that the actors continue to work from, are still conspicuously present. The actors and director are now working through scenes physically, exploring tones, colors, and possibilities *en route* to giving the scene a definite—albeit tentative—shape. The dramaturg has taken up place at the shifted table, perhaps close to the stage manager, perhaps beside the director, though the director is likely (depending on his style) to spend a good deal of this phase on his feet, moving in and out of the scene in ways the dramaturg will almost never do. As at the tail end of the table work period, the dramaturg will now assume a quieter role, and unless called upon or presented with an obvious moment for interjection into a discussion that has stopped the action and become fairly inclusive (not a hushed head to head between actor and director), she will generally make comments through the director. While some of these notes can be offered in a quick whisper during a lull in the action, others will need to be scribbled down to be presented to the director at a more opportune moment.

Those moments are still likely to come up fairly frequently at this stage, though some notes may have to wait till actors are on break, and how insistent the dramaturg is will depend on factors such as how encouraging the director is being about her spontaneous input, and how urgent the note seems to be.

Sometimes, particularly when working on a moment that has given the actors difficulty, the director will want to keep things moving whenever progress is being made, regardless of small issues or concerns. Such things can be dealt with later.

Much of the protocol of such situations will be determined by the specifics of the company, actors, and director of the given show, but some general rules might be identified. For example, while directors will often talk quietly to an individual actor, it is generally not acceptable for the dramaturg to do the same, unless she has the director's prior permission. When the director trusts the dramaturg's judgment (and that she will not offer notes which will subvert or destabilize what the director is saying), this may change, but it should not be assumed. In small matters—pronunciation, metrical emphasis, and so forth—more latitude is given, but a dramaturg working with a director for the first time should always get permission to give such notes before doing so. Directors can get understandably nervous if they see their dramaturg whispering earnestly to an actor in a corner. The note may be perfectly innocuous, but if the director doesn't know that in advance, the dramaturg can inadvertently undermine her own status and trustworthiness; in dramaturgy the former is almost entirely dependent on the latter.

Most productions that employ a dramaturg will also employ a voice coach, at least on an occasional basis, and here the dramaturg has a natural ally: someone similarly on the margins, someone whose work is crucial to the production but largely invisible to the general audience member. Having established some basic pronunciation guidelines dealing with dialect, proper nouns, whether or not to use a liquid "u" in words like "Duke," and so on, the voice coach may have been largely absent from the table work sessions, which are predominantly about thinking through possibilities and less about executing them. Once scenes start to get on their feet, the voice coach is likely to be more evident and can provide the dramaturg with useful input on matters of vocal articulation of the script, particularly in negotiating the changes made to the text at the editorial stage. While some of this may have come up at the initial read-through, other points may well arise now that the voice coach is getting to hear the script in its preliminary stages of performance. Comments from the voice coach may lead the dramaturg, after consultation with the director, to make further changes to the script, all of which should be passed on to the relevant actors and to the stage manager who is responsible for maintaining the definitive prompt copy.

In the absence of a voice coach (or one who is only intermittently present) the dramaturg may need to function in this capacity to the best of her ability. Most dramaturgs will not have the skills or training to coach an actor who needs real work, of course, but she can at least listen for issues of vocal clarity, projection (once the show has moved into the theatre), pronunciation, and metrical awareness. This last category can figure extremely prominently in these still early rehearsal stages depending on the extent to which the director is wedded to ideas about the way that scansion dictates breathing, pace, emphasis, rhythm, or even emotional impetus, in ways advocated by directors such as Peter Hall.[7] Since the dramaturg is sometimes called upon to serve as an authority on such matters, a considered position on the usefulness of metrical "clues" seems called for.

Despite the vehement proclamations of theatre practitioners who adhere to the notion that the text contains all the necessary clues for performance, I think that the degree to which actors are guided by meter and scansion is largely a matter of taste, rather than correctness. Even if we assume that the texts we have are somehow an accurate record of what actually took place on stage, it makes little sense to insist on pursuing Renaissance acting practices in this matter when modern actors (and audiences) do not replicate these practices in other matters. Since we do not hear the language in the way the early modern audience did, it seems to me perversely anachronistic to insist that actors never, for example, insert a pause that might derail the regularity of the iambic line, even in those moments of dialogue in which the ten-syllable line is broken between two different speakers. Sometimes, particularly to the ears of a modern audience, such rapid-fire exchanges need to be slowed, and that may mean disrupting the rhythm as written. There is no doubt that strict reliance on the meter can create a more fluid or more musical sound, but to insist on this at the expense of approaches that might alter the emphasis or create humanizing moments of silence seems to me perverse and reductive. That said, some theatre practitioners (Patsy Rodenburg in particular) have made excellent cases for actors re-learning to use the power and features of the Shakespearean line in order to access its unique qualities, and the obvious benefits of such an approach—if only to the actors' options—should give any dramaturg pause before tampering with the metrical arrangement of the words.[8] Certainly the disregarding of features such as metrical structure in the interests of a supposedly more immediate performance is at least as reductive as the stringent pursuit of verse-speaking. A better approach, to my mind, is one that integrates what John Barton calls "the two traditions"— verse speaking and the more modern, post-Stanislavskian focus on internal motivation—using both to create a greater range of options and effects. As a general rule, then, I would advocate the full use of the meter of the line as a starting point for rehearsal, with the proviso that it can sometimes be altered or disregarded, according to the choices of the theatrical moment.

This stage of the rehearsal process marks the point at which the dramaturg becomes less an originator—one who brings in ideas or research to help set things in motion—and more a monitor, one who assumes the role of an informed audience member who tracks what is starting to happen on stage. This monitoring of the action will be the source of much of his notes to the director and will likely focus on issues such as the extent to which the framing concepts are making sense and feel integral to the show. If they aren't, if they feel gimmicky, illogical, or inconsistent, this needs to be addressed quickly while it still can be, and it is the dramaturg's duty to raise the concern, even if the director chooses not to act on it.

Some dramaturgs believe that their ability to serve as a member of the potential audience "objectively," is best served by only occasional presence in rehearsal. I think this is questionable, not least because there is no such thing as an objective audience member, or even a truly representative one whose subjective opinion can be taken to be emblematic of the way most audience members will respond. As a rule, I think the dramaturg sees the show better from within, not from occasional visitations after which he or she pronounces comment (which is likely to be read as judgment). As Mark Bly says, the dramaturg should primarily embody a "questioning spirit," something in no way undermined by almost constant presence in rehearsal (Moore 110). Having said that, if the dramaturg cannot be—or has opted not to be—around all the time, and if this is understood by the director and the company in advance, it is at this point, with the scenes all roughly in shape and entering the polishing phase, that the dramaturg might begin to withdraw. In the case of dramaturgs who double (or function) as assistant directors, and whose role goes beyond what I am outlining here, the director may want to keep them around to help work on secondary rehearsals or do other tasks that are not properly dramaturgical. Otherwise, the dramaturg might reasonably start appearing less frequently.

RUN-THROUGHS, DRESS AND TECH.

By the time the show's constituent parts have begun to firm up adequately, it will be time to start running chunks of the show in sequence. Though much of these run-throughs will focus on production issues, the dramaturg can help by focusing on issues such as the clarity of the storytelling, or the consistency of character from scene to scene. This standing in for the audience is especially useful because everyone else involved now has a specific task of their own, which will consume all their focus and attention: as the dramaturg's role decreases, after all, the role of the designers and other elements of the company are altering inversely, becoming more focused and urgent. The director, who has to bring all the show's parts into a whole, has enough on his or her plate, so keeping a watchful eye on the show's intellectual coherence and clarity often falls to the dramaturg.

By intellectual coherence I mean not just the question of whether the production's narrative and guiding principles are comprehensible, though that is of paramount importance. I mean also the smaller constitutive elements that form the building blocks of the larger production and which have hitherto fallen outside the dramaturg's sphere. This is especially true when the run-throughs are inside the theatre and involve all the technical and design elements that have been largely conceptual to this point. The dramaturg is sometimes the only person who can sit in different parts of the theatre to take notes on sight lines, for example, or on light and sound levels; who can tell when a crucial prop is too small for its meaning to be clear; or when a character's costume is behaving contrary to intention under the stage lights. This is not simply a matter of being a spare set of eyes and ears, it is that the dramaturg is—by now—saturated in the ideas and issues behind every aspect of the show, and thus can tell when something is not functioning as had been planned, or when the mood is wrong. Yes, the director is watching for these things too, but he has so much to coordinate at this stage that not everything will get his full attention.

These issues of intellectual coherence must, again, be modulated with a genuinely collaborative sense of the show's eventual audience. The dramaturg must attempt therefore to "read" the show not simply as an "ideal" audience member (whatever that might be), but as the kinds of audience member he expects to be in attendance. Obviously such an attempt is speculative, and one should not expect uniform responses from any audience, however well one knows their demographic make up, but it can be useful to consider (if only to give the actors a heads-up) that certain moments may resonate in previously unexpected ways because of the nature of the house. When GSF stages student matinees, for example, the pace of the show tends to shift dramatically because of the ways the kids react, generating radically different colors from those anticipated in rehearsal, even more so if those audiences are dominated by certain groups (home schoolers, rural schools, conservative Christian schools, predominantly African American city schools, and so on). While a sense of audience has, hopefully, underscored everything done so far, this is the time when a clearer sense of how things might play can lead to consideration of, for example, points of incoherence or controversy.

In run-throughs, the dramaturg should sit somewhere unobtrusive in the house, where he can take notes. Notes at this stage should be presented to the director in writing, perhaps by e-mail, unless director and dramaturg have the rare opportunity to sit down (maybe over dinner or drinks at the end of the day) and chat over where things currently stand. If the dramaturg has been attending production meetings (largely as an observer), the conversation may follow such meetings. Since the dramaturg has been around since the beginning of the production, the director and actors may well come to him for a general sense of where things are. It is important to be encouraging and excited about the upcoming show, even if there are problems still to be worked out, since theatre rarely improves when those involved are suffering a crisis of faith in the production.

Enthusiasm does not, of course, preclude constructive criticism, though such criticism should emphasize what is attainable, not what is beyond saving. As Mark Bly suggests, dramaturgs should not say that a director is wrong, not because the director thinks he is infallible, but because *wrong* suggests a binary with only one correct choice. Directors, he says, generally know when something isn't working and dramaturgical notes pointing out what isn't working are powerfully annoying. Offer solutions or say "I saw this. What was your intention?" (Moore 112). Make oblique observations not attacks: "It's odd, but today I heard something else in that line that I never heard before. I wonder if there isn't another color also inhabiting that line" (Moore 113). As ever, the dramaturgical spirit is collaborative.

Throughout the rehearsal process, including the period of run-throughs and technical rehearsals, the performance script is still subject to modification. Problems sometimes emerge late. Sometimes it takes a full dress and tech.

to truly gauge the length of the show, and though it is far from desirable, drastic cuts sometimes have to be made only days before opening. Though it is better to handle such things as early as possible, it is crucial that the dramaturg recognize that such changes do sometimes have to occur at the eleventh hour, and handling such issues (director: "I need to lose a page and a half right now!") requires focus and an acceptance of the malleability of the genre. This is particularly hard when what is to be cut is something to which the dramaturg is particularly attached. In such cases, the dramaturg can offer counter arguments and suggestions, but briefly, and she should be ready to be overruled, particularly when the director feels like he is really on the clock.

The dramaturg's final function might be considered a kind of calming force, one who due to his knowledge of the show coupled with his current separation from it, is able to offer not just commentary and criticism but conviction about the value and validity of the show in ways that can soothe nerves and lessen tensions. The essentially collaborative spirit of the job can make the dramaturg a perfect counterpoint and mediator as the inevitable stress of a rapidly approaching opening night takes its toll on everyone involved. Without lights to focus, costumes to repair, fight choreography to practice, or cues to call, the dramaturg—whose low profile separates him from the director whose work is riding on everything coming together—is free to affirm, sympathize, and celebrate the art that is coming into being.

SECTION V

OPENING AND BEYOND

WRITING FOR THE AUDIENCE
(SYNOPSES AND PROGRAM NOTES)

There is something comforting about a hefty program, something that smells vaguely of polish, sophistication, and authority. These are not always good things, however, and the last is especially problematic. Programs that are crammed full of mini essays and blocks of supposedly relevant information may lend the production a certain intellectual credibility (depending, of course, on their content), but this investment in the printed words and images can have negative consequences for the production as a theatrical event even if these words and images seem well researched, eloquent, and relevant to both play and production. Before sitting down to contribute to such a program, therefore, the dramaturg needs to come to terms with certain key questions about how this particular program will work, what its aims are, and how best to achieve them.

It is not uncommon in the more lavish of theatre programs to find any or all of the following: director's notes, dramaturg's notes, a play synopsis, a biography of the author, assorted time-line charts, maps, pictures, stage history, critical commentary on the play, political quotations, essays on sociological theory, descriptions of wars or other historical events to be linked analogously to the play, manifestos of one kind or another, and so forth. All of these may be of real value to the audience. What they may also do, however, is contrary to the spirit of this book and of what I take to be at the heart of theatre itself; they reinvest the text of the program with authority at the expense of that which inheres in the theatrical event itself. In other words, such programs can potentially overdetermine the show, telling audiences how to read it, giving them answers on the page, which supposedly assure a "correct" interpretation of what will be done on stage.

Theatre, as I have said throughout this study, is neither authored by any one person (such as the play's author or the production's director), nor is it authorized solely by the texts that preexist its performances. Theatre creates art of the most ephemeral kind through the unique interaction of the conditions of the stage with the presence of the live audience. The result does not rely on textual traces, on history, or on authors to exist and be somehow true and real. To underpin all this theatrical ephemera (an ephemera that is, of course, one of the medium's greatest assets) with a program designed to steer the audience's eyes, ears, and mind is thus extremely problematic, and is why some productions prefer to function with the barest of programs, which do little more than credit the members of the company. This approach avoids the sense that the audience's reading is being weighted or prejudiced by material that we might call "extra-productional."

This last assumption is worth interrogating, however. The program is, after all, a part of the production, even though it is not part of the onstage activity, similar perhaps to the way that the physical conditions of the theatre (sight lines, acoustics, structural aesthetics, and the like) become part of the production. I am wary of the idea that the program should be banished into the world of the extra-productional, not because I want to privilege text or the position of the dramaturg, but because such a position seems faintly mystificatory, an attempt to nullify the larger material conditions of the production. Company economics, local corporate politics, audience demographics, and so forth all have a palpable effect on the show, even if they aren't clearly recognizable on stage. Theatrical production is a uniquely local phenomenon, and anything that is part of the process of that production—including the program (advertisements and all)—is legitimately, albeit indirectly, a meaningful part of that production's communicative apparatus.

If we accept that the program is a real part of the show but still want to avoid using its content to trump the onstage action, there are possibilities other than reducing the program to a mere cast list. Much of the usual program contents can, after all, be of real use to an audience and I am wary of deleting anything that can make the production richer or which can help an audience to think through some of what it raises. Part of the problem, clearly, is one of timing. The program should not, if possible, preempt the show or otherwise coerce the audience into seeing what goes on on stage through a particular lens, a contingency that might be reduced if the audience doesn't read the program until *after* the show, when it has had time to digest its less textual experience of the production. One alternative, then, might be to only distribute the program at the end of the show or (since people like to have something to read as they wait for the show to begin) to distribute only a partial program as audiences enter (containing only advertisements, a cast list, and other credits), handing out the more "dramaturgical" material (history, commentary, discursive or argumentative essays, and so on) as the audience leaves. This doesn't erase the sense of the theatre's authority being somehow punctuated by that of the text (in this case the program), but it does at least allow the production to speak for itself first.

Such approaches to the problem, however, may well prove clumsy, and audiences used to doing their reading as they wait for the lights to go down might feel that getting the program after the show is a bit like being given the menu after they have finished eating. Moreover, the dramaturg rarely controls the logistics of program distribution, and though he or she may have a say in a general discussion about what the program should contain, it is more likely—particularly in the case of the production dramaturg hired on an ad hoc basis—that she will be expected merely to supply text to fill an allocated space. This section discusses the dramaturg's options in filling that space. If the dramaturg still wishes to avoid the "preemptive" difficulties attendant on program notes, it might be worth considering the addition of the equivalent of a "spoiler alert" along the lines of: "The company recommends that you read the following essay *after* the performance."

Programs allow dramaturgs to communicate directly with the audience in a way that allows deliberation and a precision of expression not usually available in preshow lectures and the like. It is thus not surprising that dramaturgs often want as much space as possible in the program and are unwilling to give up this unusually imminent form, particularly since the nature of the dramaturgical job tends to demand that their voice is mediated and distant, at least as far as the audience is concerned. In the program, the dramaturg gets to stamp the production with his or her own voice and ideas and in a way that is lasting, a way by which the show will be remembered. Indeed, despite the concern about program notes overinflecting or determining the show, the ability to give the audience a "heads up" as to what the production is trying to do, issues or perspectives it intends to represent or engage, can be extremely helpful, and I have often noted the way that elements of such notes find their way (sometimes in direct quotation) into press reviews.[1] They can usefully announce to the audience a direction or frame of reference that can be of real value—particularly with regard to plays people know and to which they bring specific expectations that may not be in accord with the production.

There are several different ways to approach the content of the program notes (and many more that combine them, or that inhabit intermediate ground), the first of which might be that minimalist approach which has the advantage of not "textualizing" the show, of forcing the audience to watch and listen instead of relying on information printed on the pages under their seats. In fuller programs, however, the dramaturg has the opportunity to contribute a brief essay, something that can partake of several strategies. One such we might call "performance specific," as opposed to others that are "play," "concept," "period," or that "process" specific. In their purest and least adulterated forms, these approaches reflect quite different ideas about the function of the program notes.

Performance-specific notes talk about the show that the audience is about to see—or has already seen. In their most extreme (and to my mind, problematic) form, they discuss staging decisions and scene dynamics in ways that serve as commentary on or explication of the show itself, footnoting the production, as it were. I distrust such strategies, since they betray a failure to trust the communicative or discursive power of the theatrical event itself, as well as preempting that event in precisely the ways that have led some companies to pursue the more minimalist approach.

In less extremist incarnations, however, the production-specific approach assumes only that what it says should be relevant to the show the audience experiences, not to other incarnations of that show. This approach would not, therefore, include details of Renaissance history or arguments about authorship unless they had direct bearing on the production, operating instead on the assumption that the program is part of this particular and unique show and should give the present performance its full attention. An essay coming out of this tradition might discuss an idea that was central to the director's approach to the play or a theme that has come to dominate the show as it has come together. At the end of this chapter, I have included a short piece on *Twelfth Night*, which I used in a program and which might be considered performance-specific (at least in part) in that it is careful to echo tonally the mood of the production.

Play-specific notes tend to treat the play as text or at very least as a conceptual structure whose presentation on stage is separate and of only illustrative relevance. Essays of this kind quote heavily from the text or dip into literary criticism, and they are self-consciously distant from the show. At their best they provide a literary context for the production, showing the play and performance as separate though linked entities in ways that encourages the audience to scrutinize, even to challenge, the production's interpretive approach. Such essays may be of real interest and value, but they also tend to be self-marginalizing, consigning themselves to a strictly literary sense of the play, a strategy which seems curiously off the point in a theatrical program. In some cases, this sense of irrelevance can be underscored by errors of judgment or knowledge about the show itself: extensive focus on or citation of lines that the production has cut, for example. This is surprisingly common and suggests (as this overly literary sense of the play often suggests) that the dramaturg has not been extensively involved in the show at all and has been working alone, with only his folio for company. I have been arguing throughout this book that the dramaturg should be invested in the production, not merely in the text, and program essays are no exception. A purely literary essay, particularly one that ignores the focal points of the production, needs to be exceptionally good to justify its inclusion in a theatrical program, and even then it remains a sidebar that is no more connected to the audience's experience of the show than any other essay published in a literary journal. That said, play-specific essays are common in programs, sometimes because they have to be written before the show has developed in rehearsal, and they can provide a useful sense of framing or underlying ideas that only become truly problematic if they actually run contrary to the production itself. I have appended an example based on *Cymbeline*, one that draws a little on historical issues, but which remains a study of the play as literary artifact, and contains nothing that wouldn't apply if this was not a play but a novel.

Concept-specific notes will address the intellectual and aesthetic frame of the production, and it is here that the program is particularly in danger of "footnoting" the show, explaining what, presumably, should be comprehensible from the performance alone. Apart from a general statement of intent (and this is probably best coming from the director and not the dramaturg), such notes often take the form of analogous documents, quotations, and images, either as part of an essay or set alone, their relevance implied but not explained. Such uncontextualized proclamations can work well if their relevance to the production is indeed self-evident, but if the audience has to work to see the link there is surely a danger that these unexplained statements are somehow overwriting the production in questionable ways. This seems to me a clear and problematic instance of the audience's interpretation of the show being pushed (and consequently limited) by material from outside it. If such material is genuinely important to the show, it should have found expression within the show through set, costume, sound, or action. To impose such a reading on the show after the fact (and the program notes will always seem somehow *post-production* regardless of when they were actually written) is to suggest a "correct" reading of the show, which overwrites all the work that has gone into its staging, something which either undermines all that work or—more likely—renders the program and its writers irrelevant.[2] Any show that feels it needs to be explained on the pages of a document that many of the audience will read distractedly if at all, is already in serious difficulties. If a framing concept has not come across in the ways that were intended, the dramaturg should not attempt to bail the concept out in print. Rather, the show should be allowed to be what it has become, without the program second-guessing it.

One way to get around the sense of such conceptual notes forcing a reading of the show is to ensure they manifest more than one voice or perspective. A production of *Henry V*, for example, might present different images or pronouncements on one or two relevant issues—say, war and kingship. To oversell a particular agenda (the evils of war, the self-serving garnering of power under the crown) tends to oversimplify the interplay of ideas in the production and gives away an interpretive element of the show, which is no less damaging than announcing key plot points. The audience, even one that knows the play inside and out, wants to see the story and all its attendant ideas develop before their eyes. They don't want to feel that a few minutes browsing on the program before curtain has told them everything the show will reveal as it progresses.

Period-specific program notes foreground a core historical moment or moments, usually those in which the play was written or those in which the current production is set. Both can be seen as lending authority to the production, presuming of course that any Renaissance history that appears in the program is clearly relevant to the current show. If it is not—and much of the Renaissance history that appears in program notes to Shakespeare productions is not—then it is marginalia: mildly diverting or informative, but outside the current production even if it is assumed to frame that production. Such things often wind up functioning as the trappings of a cultural nostalgia or the stamp of authority, both of which tend to be unnecessary and tiresome.

The process-specific essay is one that discusses the ideas and approaches first brought to the table in the preparation of this production. This is especially useful in support of a production that began with a very exploratory rehearsal period, the initial energy being driven more by hunches and questions than by hard research. In the interest of not preempting the theatrical event, such process-specific essays need not reveal where the rehearsals led or what discoveries finally shaped the show, emphasizing instead those tentative beginnings or initial images that found their way into the show's design, those unusually resonant lines from the text or other minor epiphanies that are often at the root of a production's vision. This process-oriented writing is often manifested most clearly by director's notes, but if the director takes a different tack or the program does not include directorial comments, this may be a fruitful type of writing for the dramaturg who has been involved from the beginning.

As well as the discursive essay, the dramaturg is often called upon to write a synopsis, a subgenre, which is almost as tedious to read as it is to write. Few things undermine the authority of the theatrical event more completely than synopses—the Cliffs Notes of the theatre—which offer textual answers to the play's convolutions while giving away all the story's secrets. They mute the mystery and, worse, reward the audience's distraction and lack of focus. They make the audience passive, less collaboratively invested, offering as they do a solution to all mystery, a track from which the story can never be derailed. Sometimes a show *needs* that sense of confusion—the audience sharing the bafflement visited upon a Viola—because that confusion creates connection between house and stage, something all too easily eradicated by the program's overtidy synopsis, which stifles all suspense as it prophesies all that will happen and how. I therefore advise that where a dramaturg does have to write a synopsis, he or she keeps it as brief as possible, leaving unarticulated as much of the play's surprises as is reasonable without completely subverting the ostensible purpose of the writing. One possibility is to summarize only the first half of the play, enough to clarify the initial plot points and character relationships on which the story builds, but leaving the resolution of those plot points and relationships unstated. Such an approach would need, naturally, to be cleared by the company manager or producing artistic director in advance.

That said, anything that is written in a program's play synopsis should strive for absolute clarity if it is to be of use, particularly to the baffled audience member consulting it at intermission to figure out who someone is or what they are doing. Dramaturgs often assume that such matters are best addressed with copious detail, documenting all the characters and what is happening, but I think this is as likely to make for increased confusion. This is evidenced, for example, in the wall of proper nouns—particularly character names—which one often encounters in play synopses, something that connects primarily to the play as text, where names crucially indicate who is speaking. On stage, by contrast, one can follow the action of the play perfectly well without knowing everyone's name, and since these names occur infrequently in actual dialogue (no one ever calls the king in *Hamlet* "Claudius") they are often less than enlightening in the synopsis. Obviously some characters will have to be named, but I would keep this list to an absolute minimum, referring to the less central characters according to their relationship or plot function (the murderer, his friend, the king's daughter, her servant, and so on).

Similarly, the synopsis should track the play not in terms of acts and scenes but in terms of story, mood, emotional dynamics, or through reference to moments such as extended comic byplay. Even in the synopsis the dramaturg's duty is to the production and to keeping the audience on the narrative rails, not to documenting what happens in the play for its own sake, and lengthy description frequently crosses over into interpretation, something the dramaturg should leave to the audience as much as possible (at least as far as the synopsis is concerned). Let me give an example based on *King Lear*, beginning with a fairly detailed summary of the story.

> The play begins with a discussion between Kent and Gloucester about Gloucester's illegitimate son Edmund, which is interrupted by the entrance of King Lear who demands that his daughters tell him which of them loves him most. The prize for this love test is control of his kingdom, since the king is abdicating. Goneril and Regan use manipulative language to satisfy him, but his youngest and best loved daughter Cordelia insists on speaking frankly and explaining her love to her father as according to her bond, no more nor less, in part because she is about to be married to a man (Burgundy or France) who should have at least as much status in her affections as her father. Enraged, Lear disinherits her and gives her portion to her sisters, making Burgundy retract his suit for her hand, and leading Kent to protest his rash decision. Kent is banished for his pains, but France takes Cordelia as his (dowry-less) wife. The King says he will live alternately with his two eldest daughters, beginning with Goneril. Edmund, meanwhile has planned to revenge himself on his father for his bastardy, and does so by forging a letter from his legitimate brother, Edgar. He then pretends to be hiding it from his father who, reading its contents, banishes Edgar, convinced that his legitimate son has designs on his life. In conversation with Oswald, Goneril reveals her contempt for her father, and the fact that the King has a Fool whose companionship he especially values. Kent returns to Lear in disguise and proves his new allegiance by taking the king's side against Oswald, but Goneril expresses her irritation with the King's behavior and that of his retinue, his Fool in particular, who makes caustic remarks throughout the scene. When Goneril will not relent, Lear curses her and Albany, and vows to visit Regan instead. Lear sends Kent ahead with letters to Gloucester, and reveals through his interaction with the Fool that he is particularly afraid of going mad.

That is the first of five acts! Straw man though it clearly is, this synopsis is not substantially different from many synopses that appear in programs. I prefer something much more streamlined:

> On the eve of the betrothal of his youngest daughter, Cordelia the King decides to divide up the kingdom according to how much his daughters profess to love him. When Cordelia's answer displeases him, he disinherits her and sends her to her husband, the king of France, without a dowry, resolving to live with his other daughters. He also dismisses his loyal councilor, Kent, who returns in disguise to serve him. Meanwhile, another of his councilors, Gloucester, mistakenly turns on his son, Edgar, due to the villainy of his illegitimate son, Edmund. Gloucester does not realize Edmund's treachery until he has suffered

grievously for it and is left to wander blindly into the stormy night where he meets the now deranged King, and Edgar who is disguised as a mad beggar. Cordelia returns to take back the country from her villainous sisters but is taken in battle where she is reunited with her father, while Edgar challenges his half brother (who has become romantically involved with both of the King's older daughters) to a duel.

This significantly shorter synopsis covers the entire play, though it deliberately leaves crucial events (notably the blinding and attempted suicide of Gloucester, the outcome of his sons' duel and the deaths of Lear and Cordelia) out of the story in order to preserve their impact on stage for those who don't know what is coming. Names are kept to a minimum and relatively minor characters such as Albany and Oswald are not mentioned at all. Even the Fool (and his mysterious disappearance, which the production may consciously "solve") is not referred to because while his presence and significance are enormous, his impact on the plot itself is minimal. This synopsis offers few explanations or readings because these are not its purview. This is the most skeletal of outlines designed not only to keep the audience broadly informed as to what is happening, but to ensure that their attention remains on the staged action. Such a synopsis offers a narrative thread, and one arranged according to the most important arc of that narrative, but for detail, for mood, for explanation and for a sense of import, the audience will have to trust the performance. This, I would argue, is as it should be.

Finally, I want to add some general pointers about how the dramaturg should write for the program, regardless of *what* he or she is actually writing. Above all, essays, synopses, and any other contributions to the program should be clear and comprehensible to any intelligent and reasonably well-educated audience member, and though this (rather vague) definition will necessarily shift according to the specific demographics on which the company draws, it suggests that anything which involves an unexplained technical or professional vocabulary should be deleted or rephrased. The language of academic criticism is aimed at a professional readership, so it should not figure in a document aimed at a general audience. Here, as elsewhere, the dramaturg's task is to communicate, not to impress, and that communication should be aimed at the broadest readership possible. Again, this does not mean "dumbing down" what the dramaturg has to say, though it may mean limiting the scope of the remarks in order to express them as clearly as possible when the shared shorthand of academia is jettisoned. As my examples that follow demonstrate, I prefer a fairly chatty tone that presents its ideas in a manner which is approachable and nonthreatening. The theatre—which is where a lot of people read the program—with its milling, chattering people, its background music, its low light and atmosphere of anticipation is hardly an ideal space for reading, so the dramaturg should strive for an immediacy and clarity conducive to easy processing. As in the script, opacity is the dramaturg's enemy, and overly dense thought or phrasing will consign the program to the dark recesses under the seat.

In short, I would recommend that any essay in the program should offer a single, usable idea, and by usable I mean something that will shed pertinent light on some aspect of the play or (preferably) the production. It should do so clearly, briefly, and in ways that do not preempt the performance itself, offering not so much a reading as a relevant idea with which the reader may engage the show intellectually.

One of the problems with my continued insistence on the relevance of the program notes to the show is that many companies need to take their programs to press early in the rehearsal process or even before that process has begun. This is particularly true of repertory companies whose program covers more than one show. In such a situation, the dramaturg's first discussions with the director loom large as a way of gauging what the production will be, and while the resultant essay may finish up being more about the play than about the production, it should feel at least connected to that production and should avoid the faux pas of discussing concerns that have no part in the final show. Of course, notes written before the rehearsal process begins will always run that particular risk and there is no point worrying about it overmuch if the show veers off in an unforeseen direction. Hopefully, the printed matter in the program will become only marginal and not actually subversive.

18

19

Program Essay Examples

What follows are examples of short essays I have written for production programs to illustrate some of the distinctions discussed in the previous chapter. Naturally, since they are all by the same author and were written for the same company (GSF), they share a great deal, and consequently tend to represent those aforementioned distinctions in matters of degree rather than kind. Each one tends to present a large idea that is directly linked to the accompanying production even if that link is understated, the essay presenting the issue as being one of play as text or of historical product. To scholars they will seem familiar even generic, though I think that they are appropriate for their audience and contain points that will be of real interest—and newness—to the general theatregoer.

EXAMPLE 1: *TWELFTH NIGHT*

I cited this in the previous chapter as being performance specific, though it is so only in that it picks up the production's themes of melancholy and liquidity (the show—which was directed by Sabin Epstein in 2000—drew heavily on the play's use of water in both its imagery and its plot). It begins with some vaguely biographical throat-clearing, which had the fairly minimal virtue of linking the show to others that many of the audience had seen recently performed by the same company. This fairly predictable reflection on trends in Shakespeare's career might have been better given over to something more apposite about the show, but it at least serves as a way in to the play, and one that undermines audience disappointment that the show was more meditative and less rollicking by claiming (indirectly and somewhat disingenuously) a kind of authorial trajectory that justifies the production's overall feel. I'm not crazy about the way it anticipates the end of the play in the last sentence, but it is sufficiently vague about who manages to find love and who doesn't that I don't think it does undue damage to the experience of watching the play.

It's Better to Have Loved and Lost . . . or
"I Was Adored Once Too"

Twelfth Night was probably written in the same year that Shakespeare wrote *Hamlet* (1600–01) and marks a period of transition in the author's career.

As *Hamlet* heralded a new focus on tragedy so *Twelfth Night* marks a movement away from early comedies like *The Comedy of Errors*, *A Midsummer Night's Dream* or *Much ado About Nothing*. From here on, Shakespeare's comedies get darker, their endings less clearly satisfying, and plays like *Measure for Measure*, *Troilus and Cressida*, and *All's Well that Ends Well* (which actually ends pretty questionably) abandon the totalized comic resolution of the early plays. *Twelfth Night* stands on the cusp, a romantic, festive comedy tinged with a palpable melancholy, a play whose resolution unites some of the community with weddings and the satisfying of desire, but leaves others out, pathetic, bitter or unloved. Reality, which had been kept at arm's length in *Comedy of Errors*, Shakespeare's earlier comedy of twins and mistaken identity, now becomes unavoidably intrusive like the ever-present rain of Feste's final song.

The emblem of this sobering reality is, as it was in *Errors*, Time. But in *Twelfth Night* that focus on time has moved from the 24 hour span of *Errors*, to days, weeks, months and years. The urgency of *Errors'* action has been replaced by a languid focus on people waiting for life to deliver what they think they deserve (be it love, wealth or status), people looking back on what has been lost and trying to make sense of what remains. It is a play about love and wit and carnivalesque hedonism, but the clouds of mortality (first glimpsed in the shipwreck which seems to have drowned Sebastian and in Olivia's excessive mourning for her own dead brother) never fully dissipate. Orsino's love for Olivia is touched with melancholy which seems oddly self-indulgent as if sorrow is an end in itself. And of course, the Elizabethans were accustomed to melancholy both as a disease akin to what we might call depression, and as a fashionable pose, a piece of self-staging that showed one to be introverted and deep. But Orsino's melancholy cannot be psychologized away because it is so omnipresent in the play, from the potent image of Viola's imaginary sister who "sat like Patience on a monument, smiling at grief" to the pathos which even Sir Andrew's foolishness cannot quite conceal. It is hard not to feel some pity even for the dour killjoy Malvolio in his humiliated isolation.

This is not to say, of course, that the play isn't funny. It is, and the Malvolio letter-reading scene, for example, is as close as Shakespeare ever comes to pure farce. But the humor in the play seems to stem from strategies of avoidance, of trying to lighten what the characters privately know—or suspect—to be bleak and often pointless. Some escape by fastening onto others and finding meaning in loving relationships for whatever time they have together, but these escapees are as much the exception as the rule. By the end of the play love is, as the poet Phillip Larkin says, "still promising to solve and satisfy and set unchangeably in order," but for many it remains elusive, a dream to reach for or to recollect, to help them through the long and rainy years.

EXAMPLE 2: *CYMBELINE*

The (longer) piece that follows is roughly equivalent to what I call a play-specific program note in that what it has to say is rooted not in the production or in relevant contextualizing details or ideologies, but in the play as book. As such it resembles that (now somewhat antiquated) "fire-side chat" mode of criticism, its voice at once avuncular and a little pretentious, its concerns vast and overarching: nationalism, gender politics, aesthetic structure (drama-turgy in the other sense of the word) all surface here. In the end it is some-thing of a defense of the play itself, though in the context of the production for which it was written (directed by Nancy Keystone in 2003), it was also a way of announcing a certain surreal quality that the show embraced from the outset and had done since before rehearsals began (the essay had to be writ-ten before the actors met with the director). As such, the essay that seems so textually oriented is also linked to the production, albeit tacitly, and there is thus a sense that the textual reading is being used to affirm the authenticity of the show: as ever, this is a questionable strategy, but one that audiences often find comforting.

Cymbeline's Dream Mirror

Cymbeline is an odd play. Even compared to the other, more familiar romances which characterize the end of Shakespeare's career (The Tempest or The Winter's Tale) it's odd, though figuring out what makes it odd might help us get a better handle on it.

All the romances are odd in that their plots tend to hinge on fanciful, quasi-supernatural coincidences and magical interventions, their characters frequently revealing an innate nobility of blood that no amount of unsophisticated nurture can mask. Indeed, the romances often feel less like the Shakespeare of Hamlet than they do like the brothers Grimm or Hans Christian Anderson. The romances smack of fairy story, complete with wicked stepmothers and that Princess-and-the-pea logic that says that true royalty will always shine through, as if to suggest a natural hierarchy inherent in society.

Even in such a context, however, Cymbeline is odd. Much of this stems from its hybrid quality, the sense that the author drew from several of his earlier plays in its construction. Its preoccupation with English/British nationhood is straight out of the history plays, though its actual setting is that largely mytho-logical ancient Britain reminiscent of King Lear. That Britain is vying for inde-pendence (in curiously inconclusive fashion) from Imperial Rome, and that much of the play's tone in matters of philosophy and the supernatural seems so directly linked to Julius Caesar, Coriolanus or Antony and Cleopatra has led critics to refer to Cymbeline as Shakespeare's last Roman play. The test of Imogen's fidelity seems akin to a problem play like Measure for Measure, and the cross-dressing of the heroine cut off from her beloved is a throwback to romantic comedies from ten years earlier, plays like As You Like It and Twelfth Night. The compositors of the 1623 Folio (the earliest text of this play) grouped it with the tragedies—the only genre to which it clearly doesn't belong—a desperate gesture which announces pretty clearly that they had no idea what to do with it either.

Of course, Shakespeare is (in)famous for his mingling of tonally disparate plot elements, so Cymbeline's tendency to have a foot in every camp shouldn't be that disorienting. What makes it so, I think, is that those different elements co-exist less easily here than they do elsewhere in Shakespeare. Yes, Macbeth has the light relief of the Porter's scene, but we would never consider the play a comedy. Much Ado deals with life and death matters tied up to questions of female chastity, but there is never any doubt that things will be resolved.

Cymbeline is odd because no single tone dominates the story and its competing elements often seem to be mutually antithetical. Consider the nationalism issue, for example. The English defiance of Roman rule is cloaked in rhetoric that wouldn't be out of place in the Henriad, all noble ancestry, natural bravery and island fortress stuff. But those speeches are all in the mouths of the "wicked queen" and her idiot son, and at the end of the play they are dismissed as the victorious British *choose* to become once more a tribute-paying satellite of Rome; it's tough to imagine Henry V making similar overtures to the French. Even so, Cymbeline's capitulation to Rome at the end of the play might have been less bewildering if so much had not been made of the noble and courageous British defiance of the Romans.

The key, perhaps, is in the play's fairy tale trappings which suggest a different notion of what the play is trying to do. Hamlet talks about drama striving "to hold as 't were the mirror up to nature," a conception of theatre which is basically realist. *Cymbeline*'s mirror, however, functions like the mirrors one might encounter in dreams, and the reflections they cast are shadowy or distorted, possibly *true*, but not *real*. We are not in conventional reality in *Cymbeline*, and what seems like political paradox (the British shaking off Roman rule only to take it on willingly) makes more sense if seen less as statecraft and more as a species of dream-mirror in which things that seem to be opposed to each other are shown to be somehow the same. In front of the play's dream-mirror, everything that makes things seem different falls away, leaving a strange blurring that underscores sameness in what is reflected. The British accept Rome because Rome stands for certain things the British value (the Elizabethan/Jacobean British in particular), things like self-restraint, decorum, order, religion, and, of course, imperialist expansion. Cymbeline's Britain can accept the yoke of Rome having first proved its ability to throw it off because in the dream-mirror, Britain *is* Rome. The dream mirror at the end of the play reflects not a historical reality but an impressionistic closure in which Britain's Roman-ness shines through.

The dream-mirror sets most of the play in Wales, but it is a Wales without Welshmen (think of Fluellan in *Henry V* or Evans in *Merry Wives*). Here, Wales has become strangely Anglicized, tied to English resistance of invaders and made a seat of British identity. As in the mythically Welsh origins of Arthur and the more recent origins of the Tudor line, Wales here has become an instance of the kind of cultural imperialism practiced by Rome. Though once thought barbaric, the Wales of the play has been absorbed into Britain, if not into England, and as such the dream-mirror shows it to stand for other parts of the British isles—Scotland and Ireland—whose absorption into a British nation was one of King James' pet projects. In reality such absorption did not happen for many years (and then happened in ways that revealed more difference than sameness), but in the play's dream-mirror, the wish is fulfilled and Britain's disparate parts are harmoniously united in ways even Henry V could not achieve.

It might also be said that the final images of unity reflected in the dream-mirror which is *Cymbeline* are all male. The dangerous queen (who is clearly rooted in ancient archetypes of female rebellion such as Boadicea [or Bonduca]) is conveniently eradicated. Shakespeare's historical sources were wary of queens, even nationalistically defiant ones, because their very authority was seen to be a usurpation of male prerogative. The Romanization of Britain at play's end thus evokes a masculine world where women disappear so as not to trouble the natural order of things. The only woman left on stage (Belarius's dead wife Euriphile is one of many absent mothers in Shakespeare) is Imogen, a woman who was of vital importance to Posthumous as an ideal of chastity, but who is now dressed as a boy (and, of course, was originally played by a boy) and stays in her boy's garb till the end. The dream-mirror magics the unnerving femininity away, so that Posthumous' earlier misogynistic wish that men could somehow give birth without the taint of womankind seems to have come true.

A less cynical reading might track Posthumous differently, but even here the dream-mirror produces a kind of clarity. Early in the play, the dream-mirror reveals that the supposedly noble and chivalric Posthumous is bound to the imbecilic and brutal Cloten, their sameness revealed in their shared attire but rooted primarily in a limited sense of women.[1] In the dream-logic of the play, Posthumous has to become a more idealized version of himself, something that requires the killing of his alter ego. One of the play's oddest moments in terms of realist story-telling (that in which Imogen wakes with the headless corpse of Cloten and takes the body for Posthumous) thus makes sense, but it is a sense based in that deeper, impressionistic logic that is the stuff of dreams.

There is truth here, but it is that "down the rabbit hole" truth where you shed reality as you fall. *Cymbeline* is an odd play, but its oddity is not only consistent, it is—paradoxically—both the core of the play's power and a window onto the fictional nature of art in general. Since all art is structured, logically and aesthetically, its distance from reality is always greater than it may seem, however "realist" the representation. *Cymbeline* is, if nothing else, honest about its own unreality and if we who are used to valuing the semblance of realism above all things in drama find it odd, that is, perhaps, our problem.

EXAMPLE 3A AND 3B: *MACBETH*

The short essay that follows is one of those historically specific fragments of a program that had absolutely nothing to do with the performance itself but which I wrote out of a vague sense of scholarly duty. It was written early in my dramaturgical career, before I had realized the import of talking about things that were clearly part of the production (dir. Briggs, 1998) rather than merely a part of the play as textual/historical artifact, and for a production from which, for various reasons, I felt largely shut out. The result is possibly interesting (though familiar to anyone who knows anything about the play in its original cultural context) but clearly marginal in terms of the performed event. Indeed, I find it hard to conceive of a production that could seriously engage with the issues of kingship I present in the essay, they are so alien to most modern people's sense of self and government. The essay's redeeming feature is that it is ultimately about how we tend to oversimplify the past, and that it is therefore as much *about* history as it is itself history. This, I think, has some value as an engagement with a relevant idea, but its degree of remove from the production remains almost immeasurable, and I offer this less as an example to be followed than as a cautionary tale, an unfortunate enactment of its own final metaphor. The subject matter presented here might be better suited to a preshow lecture, the history serving as a springboard into a discussion of the play's engagement with politics.

The second example (3B) came from a more recent production (dir. Fraches, 2004) and though it still deals in part with specifically historical and textual issues, its approach to the play's politics is more clearly relevant to the production the audience actually saw. It speaks, moreover, to political issues whose currency is not obscured by the more archaic interest in kingship, and since its ideas seemed to me relevant both to the production and to the cultural moment in which that production was staged (with the country at war and elections looming), I think this second approach to the play is better than the first, not just different.

Our Royal Master's Murthered: Killing Kings on Shakespeare's Stage

"The state of monarchy is the supremest thing on earth; for kings are not only God's lieutenants on earth, but even by God himself they are called gods." James I

If one had to make a statement about what people in twentieth century America thought about any serious issue, it would be almost impossible to produce anything that was both accurate and useful. We are too diverse a group to hold much in common at any given moment. Nevertheless, while the complexity of our society and culture is alive to us now and gives us pause before making rash generalizations, we are quick to make those same generalizations about the world in which Shakespeare lived. All too frequently people confronted with an issue in one of Shakespeare's plays (be they scholars, teachers, students, directors or audience members) fall back on the old fiction: *that's what they thought "back then."* But who is the "they" and when exactly?

One of the issues most likely to produce one of these blanket (and dismissive) statements, concerns royalty. But what Shakespeare's audience thought about their monarch and monarchy in general is as complex a question as what the audience of this play thinks about the presidency of the United States. Yes, the early seventeenth century was used to the notion of royalty and most people were probably unused to any other notion of political structure, but that does not mean that they bought the official rhetoric about the King as the right hand of God. Nor were they unfamiliar with the idea that the killing of a legitimate King or Queen might be morally acceptable. In the religious turmoil of the sixteenth century, Protestants and Catholics alike had both developed notions of lawful regicide. A few years before *Macbeth* was written (c. 1605–7), the mother of King James, herself then Queen of Scotland, had been executed by Queen Elizabeth for supposed involvement in Catholic conspiracies. James' own son, Charles I, would be formally tried and executed by his people in 1649. Rather than seeing James' absolutist claims for the divinity of Kings as indicating what "the people" believed, then, we might perhaps see such claims as an attempt to bolster an idea to which there were significant pockets of ambivalence.

Shakespeare had killed Kings on stage before, and we have evidence to suggest that the authorities of the day were not altogether happy about it, though it probably depended on which kings were killed. Shakespeare's depiction of the deposition and murder of Richard II had drawn censorship, perhaps because Elizabeth saw something of her position in Richard's. Richard the *Third*, however, was not only a "bad" king, he was, in official history, the monster whose death gave rise to the glorious Tudor and Stuart dynasties, so his death in battle at the hands of Elizabeth's grandfather seems to have caused no such outrage. The problem for James, who traced his lineage to *Macbeth*'s Banquo, is that the notion of an absolutist monarchy in which the King is a god, makes no distinction between good and bad kings. The monarch is untouchable and a tyrant has to be borne quietly by his people until God chooses to do something about him. From James' extremist standpoint, the killing of Macbeth (whose coronation is legitimate according to the *electoral* mechanisms of Scotland's monarchy) is no less troubling than the murder of Duncan.

Assessing Shakespeare's position is, as ever, tricky. He whitewashes the historical accounts of both Duncan and Banquo (the former being no saint and the latter being involved in his murder), and gives his audience moral opposites to Macbeth in both the King who begins the play and in Banquo, his loyal thane. But even as the play makes certain connections to James himself (notably the Porter's references to the "equivocating" testimony of Jesuits involved in the 1605 Gunpowder Plot), it is difficult to turn the play into Shakespeare's compliment to King James which it is sometimes held to be. Apart from the fact that Shakespeare could have chosen subject matter that would have cast his sovereign's lineage in better light, the play *as play* belies the moral binaries often applied to it. Yes, Macbeth and his wife behave with increasing brutality, but as characters they are the center of the play's life, energy, even sympathy. On the other hand, the forces of light and goodness are not without their flaws: Duncan is politically clueless, Macduff abandons his family, Banquo keeps his guesses about Macbeth to himself, and all future hopes are pinned on the inept and strangely duplicitous Malcom. It is Malcom, not Fleance, James' forefather, who takes the reins in the last act, a fact that leaves the question of Banquo's line of Kings enigmatically unresolved.

Like those murderous Jesuits of 1605, the play itself equivocates, from the misleading riddles and half-truths of the weird sisters to the tortuous syntax and paradoxical phrasing ("fair is foul and foul is fair") that characterizes the protagonist's most famous speeches. The play's ambiguities and uncertainties infect almost every line so that nothing, be it the definition of masculinity or femininity, the natural world, or the basic laws of friendship, kinship and hospitality, can be considered reliable or stable. In this play everything, and that includes those ideas of Kingship which Shakespeare's audience are sometimes thought to have accepted so unquestioningly, is subject not just to change, but to inversion. As is often the case in Shakespearean tragedy, then, the restoration of order at the final curtain is largely a hopeful illusion, the playwright leaving more than enough loose ends to entangle the future.

What Bloody Man Is That?

The first thing we hear about Macbeth is that in the course of the opening battle against the rebel forces, he cut his way through the enemy until he faced Macdonwald and "unseamed him from the nave to the chops" (1.1.22). This graphically violent image of splitting someone open from stomach to throat reappears in different terms in the final scene of the play when we learn that Macduff was "from his mother's womb untimely ripped" (5.8.15–16), was, in other words, born by Cesarean section. The play thus pointedly begins and ends with a version of the same offstage horror.

These two unseamings are, of course, different. One, for the purposes of the play, begins Macbeth's political and murderous rise to power since it is his ferocious loyalty to Duncan in battle which leads to his being made Thane of Cawdor. The other is the way Macduff enters the world and undercuts Macbeth's sense of invulnerability. In other words, the first unseeming gives birth to a monster, the second to a hero.

But there is more to it than that. In a play so obviously anxious about issues of gender difference, the manner of Macduff's birth seems to enact a familiar archetypal fantasy in which men somehow avoid the touch of the feminine altogether. The idea is an old one, enshrined in folk tales and versions of Christian theology: women are the root of sin, corruption and weakness. This is why in Malcom's frankly bizarre testing of Macduff, he affirms his own innocence by claiming that he is as yet "unknown to woman." According to this version of women then, the true hero (like Christ himself, according to this version of the myth) should have no biological mother and has to be born miraculously. For Macduff, this birthing takes the form of being removed from the body of a mother who would be dying or already dead (Medieval and Renaissance medicine had no way of compensating for the blood loss resultant from a C section). He emerges into the world somehow untainted by the feminine and this is in part the source of his magical power over Macbeth and of his generally heroic status.

As is usually the case in Shakespeare, however, it's not that simple, and the play's final attitude to women is a good deal more complex than that folk tale element would suggest. Yes, the witches are in league with dark forces, but a good deal of their strangeness comes not from their femininity but from the odd way in which they straddle categories: "You should be women," says Macbeth, "and yet your beards forbid me to interpret that you are so" (1.3.45–6). Yes, Lady Macbeth is partly responsible for Macbeth's murderous ambition, but she is able to do so because she performs—or attempts to perform—an "unsexing" ritual when she receives her husband's letter, a divesting of her biological femininity so that her purpose will not be undermined by her female "nature." It is this defeminized version of herself which she uses to fuel her husband's killing of Duncan, though it seems—significantly—unsustainable, and she relapses into more conventional images of femininity: hostess, supportive wife, and finally the alienated and tortured mind shut out of all issues of power and control.

Indeed, the only unproblematically *female* woman in the play is Lady Macduff, and while she seems to fulfill the contemporary expectations of a woman (she stays home with the kids), she is shown not only to defy the old myth of tainted womankind (being, apparently, untouched by evil), but to defy also the idea that her husband is in fact a hero at all. She points out that her husband left his family within Macbeth's reach while he worked to build a military solution to Scotland's problems. He lacks, she says "the natural touch." Failing to stay and guard his family is "unnatural," a violation of manhood as great as Lady Macbeth's failed attempt to violate her own femininity.

The problem is that for everyone else in this play, manhood is not so much about being a good husband and father as it is a propensity to and facility for violence. It is because of this that Lady Macbeth is able to say to her husband that when he was prepared to kill his king "then you were a man" (1.7.50). Macbeth proves his manhood to Duncan by killing Duncan's enemies and proves it to his wife by killing Duncan. Macduff, in turn, proves his manhood by killing Macbeth.

Despite the misogynist myths underlying the culture of the play, the inherent problem in Shakespeare's version of Scotland is its inherently violent notion of masculinity. The unseaming which gives birth to Macduff, then, announces not so much a magical heroism, as it does a continuation of a violent cycle which pointedly excludes the feminine. It is perhaps not surprising, that the final curtain does not bring total closure: the hero who killed Macbeth is not on the throne, nor is Fleance—the son of Banquo who the prophesies suggested would be king—and we are left to wonder what kind of rule will be enacted by the shadowy Malcom. The only thing we can be sure of at the end of the play is that the getting and holding of power will remain in male hands, and in this world those hands are always bloody.

EXAMPLE 4: *JULIUS CAESAR*

The production for which these notes were written was, as I have said earlier, set in 1930s Louisiana, the play's title character being modeled loosely on Hewey Long (dir. Dillon, 2001). The director's notes addressed this "adaptive" strategy, and other parts of the program contained snippets of information and images reinforcing the link between the play and the more modern setting by emphasizing the latter. My short essay thus did nothing to make that tie explicit, though it was careful to support the nature of that tie as a principle, making a general argument connecting the politics of the play to more recent political dynamics. The agenda was to make an intellectual case for a production that some may have considered gimmicky, but to do so indirectly and from the position of an academic response to the play text and its origins in history.

Political Theatre/Theatrical Politics

One of the dominant concerns of *Julius Caesar* is what we might today call "spin:" the way politicians (and their media representatives) inflect or color events so that they will be favorably received by the public. As such, and as my pushme–pullyou title suggests, Shakespeare's play is not simply a piece of theatre with political content, it exposes the intrinsically theatrical nature of politics. Both in theatre and in politics, the principle actors survive only as long as their respective audiences permit them to do so, and as any performance on stage can be wrecked by booing and heckling, so any politician knows that his or her time in the public eye will last only as long as he or she preserves a positive rapport with the people. In *Julius Caesar*, acquiring, wielding and losing power all hinge on issues of public image as much as they do in *Spin City* or *Wag the Dog*.

Caesar is a performer and, for the most part, his performances produce the desired result: popular adoration. But while his performance of unwavering certainty in his own rightness wins him the respect of the masses, that same performance (and the extent to which it *is* a performance is exposed by Caesar's many poorly hidden frailties) can also get him damned as a tyrant. Caesar clearly thinks of himself in third person (*what would "Caesar" do?*) whenever he is in doubt, but the resultant performances of independent will and personal strength ("I am constant as the Northern Star," etc.) ironically play into the hands of the conspirators. Thus, when his enemies come to charge him with being autocratic and he responds (as third person "Caesar") by being autocratic, he confirms their complaint and leaves room for no further discussion. Casca's "Speak, hands, for me!" as he puts the first dagger in, nicely illustrates the way that Caesar's attempt to show that he is deaf to all petitions, and is, therefore, a strong and admirable leader, has backfired.

In other words, Caesar runs into one of the great problems of the theatre; audiences do not always "read" an actor in the way he intended. People at a show make their own assessment of why a character does what he does, regardless of the actor's "motivation," and sometimes they seem to read quite against the grain of the performance. As Cicero says "men may construe things after their fashion, clean from the purpose of the things themselves" (1.3.34–5). Watching a show is a subjective business, and it is neither appropriate nor helpful to have an actor come to you after the show and try to demonstrate that you misread his character. The audience member can always point out that whatever the actor had *tried* to do, it didn't "play."

Even good actors get into difficulties when they find that their earnest motivations do not necessarily persuade an audience. Brutus thinks that the image of himself and the other conspirators pacing the streets of Rome after the assassination with their arms literally dripping blood will show the people that they have been "sacrificers" rather than "murderers." But once Antony has had his way with the people, all those bloody hands seem merely to confirm their guilt and brutality. The piece of political theatre backfires because Brutus doesn't know what every actor does: that the intentions behind your performance cannot guarantee the "correct" audience response.

When Mark Antony sees Octavius's servant weeping over Caesar's body and promptly feels tears start to his own eyes, he learns two powerful theatrical truths. First, the power of visual spectacle: Brutus can talk as long and as cleverly as he likes, but the sight of the bloody corpse may well wipe out everything he has said if suitably framed. Second, the best way to produce emotion in another is by showing it in yourself: "Passion, I see," he says, "is catching, for mine eyes, seeing those beads of sorrow stand in thine, began to water." In other words, an actor generates feeling in his audience by showing himself to be emotionally overwrought. Not surprisingly then, mid-way through the funeral oration, Antony breaks down—or seems to—and the audience that had cheered for Brutus moments before, now start bellowing for his blood.

The Elizabethans knew all too well that politics was theatre. The Queen herself carefully regulated her own public appearances (in reality or in pictures) so that they would produce the right response in the audience. She encouraged the fetishizing of her Virgin Queen status, she insisted on the grandest or most sensual clothes, she appeared in armour to counter her misogynist critics, and she ruled her court with a very public rod of iron. When, in the year *Julius Caesar* was written, her highly popular favorite Robert Devereux, 2nd Earl of Essex, reached for his sword during a heated argument with the queen, she had him placed under house arrest, after which he led an open revolt against her. The night before Essex's rebellion, his supporters paid Shakespeare's company to stage *Richard II*, a play about the deposition of a king, apparently to whip up audience sympathy for their cause, recognizing that even the literal theatre could play a part in politics where politics was so theatrical. They appear to have made Brutus' error, however, because *Richard II* is a complex play, one that garners as much sympathy for the usurped king as it does for the rebels. The rebellion fell almost comically flat, and Essex was executed.

Brutus ought to have learned early in the play to distrust his own ability to read events. When he hears the crowd cheering Caesar off stage, he assumes (because this is what he fears) that they are offering Caesar the crown. In fact, as we learn afterwards, they were cheering because Caesar had *refused* the crown. A cheer by itself means little and is open to a thousand interpretations just as blood by itself can mean murder, sacrifice or other things depending on the circumstances. Brutus makes many mistakes in the course of the play, but none are greater than these two: the mistaken assumption that he can read truth from signs (that Antony will be no threat after Caesar's death, for example) and that signs can be manipulated to communicate precisely what you want them to (bloody hands = purgers of the state). But as Cicero points out, signs are open to competing readings and misreadings. As the grand arena of communication through signs (actors, scenery, light, music, etc.) theatre is thus supremely unstable. It lends itself well to political spin, but anything which spins is difficult, perhaps even impossible, for a single person to control. Antony, by contrast, knows how to do enough to increase the speed and wildness of the spin but is sufficiently politically and theatrically savvy to know when to get off stage. So, in the funeral oration, once he has given his own spin to the assassination and to Brutus' words ("honorable men" for instance), he relinquishes control of the entropic torrent he has created with the simple phrase, "Now let it work." A good politician, like a good actor, knows how to manipulate an audience, but a great politician (or actor) also knows when to stop.

EXAMPLE 5: *A MIDSUMMER NIGHT'S DREAM*

The following short essay draws briefly on the play's performance history in order to make an argument about a crucially theatrical dimension to the play, one that the current production was making use of, namely, doubling. The result, again, offers a tacit justification and clarification of a practice inherent in the production through recourse to both history and textual study. The production for which it was written was directed by Richard Garner in 2000.

Dream's Dark Magic: Desire and Self-Discovery in the Forest

"The most insipid ridiculous play that ever I saw in my life." Samuel Pepys, 1662.

One of the paradoxes about *A Midsummer Night's Dream* is that its immense popularity on the stage was, for many years, rooted in elements that had little to do with the content of Shakespeare's play itself and more to do with the ways in which the play had been embroidered with graceful music and balletic movement. Even when Shakespeare's text was largely restored (the eighteenth century had reduced the play to a series of musical interludes and dances), the emphasis remained on the operatic, the balletic and the visually spectacular. The result was that the play tended to be staged as beautifully graceful but airy to the point of being vacuous: a delightful piece of fluff. It was precisely such a notion of the play (and of the staid, remote nature of Shakespeare in general) that was targeted by Peter Brook's landmark production in 1970. Instead of a quaintly realist set populated by gauzy, dancing fairies, Brook's famous white box set, trapeze and intimate, stripped-down action, sought to reconnect the play to the present and to the audience. The production avoided the play's conventional aesthetic for a darker, more energized and explicitly erotic approach to the text, and its success revolutionized the staging of Shakespeare. Despite this success, however, *Dream* more than any other Shakespeare play, often seems to fall back on that Victorian heritage when staged today, as if audiences and directors want to keep those darker and more immediate strands of the play at arms' length. Even when the humor of the play-within-the-play is allowed an earthier dimension, modern productions still drift towards the vapidly pretty and sanitized air of the nineteenth century.

Some of this is, no doubt, due to the explicitly unreal nature of the play and the dream-like feel of the forest. Shakespeare is fond of forests as places of magic and discovery (think, for example, of the contrivances facilitated by the forest of Arden in *As You Like It*), but it is only much later in his career that they become as explicitly enchanted as the one outside Athens, and only in one of these, *The Tempest*, do we meet creatures such as those in *Dream*. It is worth asking, then, why he enters such obviously preposterous territory as fairies and magic potions. While we might (somewhat patronizingly) argue that the Elizabethans were more likely to believe in such things, we might be better served by accepting the play's supernatural elements as the stuff of dreams and of theatre. The forest and the fairies exist to resolve tensions clearly apparent in the play's first scene, to effect closure and resolution not in the pseudo-realist manner of the TV sitcom (in which all the problems of life can be neatly resolved in half an hour minus commercials), but by reminding the audience that the theatre is itself a magical space and that what takes place on its stage reflects reality through a distorting glass.

This distortion, however, does not make the content of the play any less intense or applicable to reality, as Brook understood. The play begins with a theme (forced marriage) which could easily turn tragic, as it does in *Romeo and Juliet* which Shakespeare wrote at virtually the same time and which *Dream's* play-within-the-play (*Pyramus and Thisbe*) seems to parody so clearly. For the comic resolution to be truly satisfying we have to see the tensions of the first scene resolved, even if some of those involved (Theseus and Hippolyta) are only on stage at the beginning and end of the play. Their absence from the play, however, does not separate them from its action, since what takes place in the forest clearly has a bearing on how their own relationship will stand at play's end. It should be remembered that in Shakespeare's day, many of the actors would have played more than one part, and I think that *Dream* is a rare instance in which modern productions almost have to follow this practice in order to make sense of play. Doubling allows people from the first scene, for example, to take on a different identity in the forest and do things that in reality they could not. So in our production, Theseus becomes Oberon, as if the Duke's dreaming mind has given him an alter ego. We are used to the idea that problems get clearer when we sleep on them: we wake up with ideas or solutions as if our brains have been working as we slept. The play literalizes that idea, allowing us to see the dreamscape created by the characters as they work through the strange, scary, playful, erratic (and erotic) logic of dreams, moving toward a psychologically appropriate resolution.

What the forest and the fairies represent to me, then, is a fantasy of wish fulfillment and experimentation, in which problems are solved by pushing the envelope of experience. The problems of the first scene center on love and the power dynamics that love produces, particularly where one person is dominant or where one's love is rejected. In the dream-like world of the fairies, lovers can change partners to clarify their desires, and those who had clung to power-positions can see what happens when that power is abandoned or, in Oberon's case, scrutinized by being exaggerated. In the enchanted wood, people get to vent their anger and confusion and desire, shifting their affections, submerging themselves in the utter powerlessness that love can produce and, in the process, are given a clearer perspective on their waking lives and their actual relationships. Shakespeare's fondness for letting his characters engage in role-playing here becomes literally true to an extent that they risk a loss of self and sanity as they explore the possibilities of love and desire which the woodland magic permits. The forest, in other words, allows those in it (and I include the audience) to play out fantasies which in reality might be too dangerous to acknowledge, let alone to indulge.

20

TALKING TO THE AUDIENCE

Much of what I have said about writing for the audience is equally applicable to matters of talking to the audience, not least of which is that issue of exactly *when* the dramaturg's offerings are to be processed. While dramaturgs are sometimes called upon to talk to audiences (or potential audiences) in settings quite separate from the theatre, the bulk of dramaturgical lectures and talkbacks take place either immediately before or immediately after a production. Unlike a program, which can be read at any time before, after, or during a show, the temporally specific nature of a lecture gives the dramaturg a clearer sense of what is and is not appropriate content for her talk. As a general principle based on the concerns of preempting the show outlined earlier, the dramaturg should talk largely about the *play* in preshow lectures (though not in terms of plot summary), and about the *production* in postshow lectures.

Many dramaturgs with years of college classroom experience find preshow lectures uniquely difficult because the audience can be so incredibly varied: graduate students writing rigorously historicized Marxist analyses of the play, high school teachers who teach it every year, spouses dragged along for company's sake, people who have never read a Shakespeare play or set foot in a theatre. How does one pitch a lecture to such a disparate crowd, some of whom want a plot summary, some of whom are ready to cross-examine the company with an arsenal of quotations from the play and a vocabulary filched from Derrida or Bourdieu?

First, I would say that the dramaturg needs to be prepared to talk about a core issue or idea for about two-third of the allotted time, and should be equipped with notes and a quotable piece or two of text. I prefer to keep the mood informal and to use my notes to give my remarks a loose structure while maintaining a conversational tone: a formal lecture read at a podium seems unduly stiff as a preamble to a theatrical event. This air of informality need not be undercut by knowing roughly what you want to say and how you are going to say it, indeed I think this shows respect for the audience. People who have taken the time to come to the theatre early, generally want to feel that they got something substantial out of the experience, not that the time was taken up by someone who improvised the whole thing or allowed it to be steered entirely by audience questions. Most of us are not that good improvisers, and audience questions can dictate an exchange in which many of those present have little interest, so some planning seems in order.

As far as gauging the level at which to talk, tone is all. As a rule, I aim high in terms of content (rich or complex ideas and observations relevant to the text or to the historical moment that produced it) but do my best to express those ideas in clear, everyday prose. I don't believe the preshow lecture's purpose is to explain the play by means of synopsis, and I am resigned to the fact that some of those who came expecting that may get lost in the course of my talk, though I am happy to make room for more basic questions after the more scripted part of the session is over. I do not think the dramaturg should talk about set, costume, or staging choices in the preshow talk, or give away anything about the production's strategies of interpretation and construction, or about the story, unless it is clear that everyone present has read the play in advance.[1] Rather, I would recommend taking a strategy similar to what I advocated in program note essays, addressing issues linked (however peripherally) to the play or its period. The preshow lecture, however, unlike writing for the program, allows the dramaturg to base what she says on a production she now knows, not one she had to anticipate before rehearsals had begun. This allows the talk to be both more specific and more incidental, the speaker being able to select a subject that is interesting of itself but which will be seen to resonate more fully *after* the audience has seen the ensuing performance.

Talkback sessions *after* the show are bound to be more about the production itself, which is as it should be, and they frequently involve some of the actors and even the director. In such a situation the dramaturg needs to resign herself to the fact that, as far as the audience is concerned—fresh from their experience of the staged event—the dramaturg is the least interesting person on stage. They want to hear from the actors and they want to hear from the director. They don't, for the most part, want to hear from the dramaturg, and she is likely to be asked to speak only when a question comes up that no one else wants to answer. While the director will be expected to field comments on basic interpretation and concept, the actors on individual choices, their sense of their characters, and so forth, the dramaturg will be left to deal with matters of history, text, and author. These can all bog the proceedings down and get away from the understandably sexier, more obviously creative allure of the cast, so the dramaturg needs to learn how to handle such queries quickly, particularly those that can open huge cans of worms ("Did Shakespeare really intend this interpretation?" "Did Shakespeare really write this?" "What did the Elizabethans think about women/God/war?"). Such questions are, of course, perfectly valid, but they can easily hijack a discussion most people present want to be about the show: my answers to the previous listed questions are respectively "it doesn't matter," "yes," and "lots of things," each fleshed out with no more than one or two kindly but decisive sentences. If individuals want to discuss such things further, you can always chat after the larger group has broken up.

There is no question that part of the dramaturg's visible presence at talkback sessions and lectures is to function as the company's scholarly representative in ways intended to accrue status to the company and under-write the production with a sense of legitimating authority. It is easy to be cynical about such a function, particularly if you adhere to the belief that such notions of authority are largely spurious, but such a view can easily detract from what is a genuine pedagogical opportunity. However much I don't think that something intended to be about a production should be shang-haied by the dramaturg's ideological hobbyhorse, I do think it perfectly rea-sonable for the dramaturg to use her authority and the production to address wrongheaded principles, misconceptions, or ideas that need holding up to a more rigorously theoretical light. Again, this is part of the dramaturg's col-laboratively educational function, and again, tone is key. She should avoid looking like she's on a soapbox because the audience didn't come to be preached at, but should feel free to avail herself of the chance to address, for example, why productions done in Renaissance dress are not automatically better than those that aren't, or why it is not just acceptable but essential that the Shakespearean script is edited for performance, or that there is no such thing as a "straight" or "pure" production. These, and issues like them, need to be addressed in order to produce an audience that is better prepared to appreciate more interesting, challenging, or demanding productions. Talking about such concerns under the valance of talking about the production is part of the dramaturg's educational duty, and while the consequences of such talks are small (like most of the measurable results of the dramaturg's work), they can create ripples that expand throughout the culture producing, one hopes, a more receptive environment for future productions. Since the production requires an audience to create meaning theatrically, it is worth taking the time to work with that audience as part of a larger collaborative project. The company and the audience are, after all, part of an intellectual community that crucially contextualizes this and future shows, defines them even: investing in an audience's intellectual development is an investment on behalf of the company and the community itself.

One last thing about talking to audiences: theatrical communication goes both ways, and the dramaturg's collaborative approach to audiences means learning from them as well as teaching them. Here as elsewhere, then, the dramaturg needs to listen, and not just to gauge when people aren't under-standing him. People so often confuse the theatre with something like watch-ing a movie that they forget how much audience response matters. Other than box office returns, newspaper reviews, and occasional letters of outraged complaint (which are rarely representative of most audience members' feel-ings), the company hears little about what audiences think of what is going on on stage, what they would like to see more of, what they find stimulating, and so on. Since audiences are rarely starstruck by the presence of a dra-maturg, talkbacks and lectures provide a rare opportunity to hear such things in fairly unvarnished form, some of which the dramaturg might learn from or feel are worth passing on to the director or company manager. Some of what the dramaturg hears will need addressing because it is based on faulty logic or misconception—though even that may indicate a need for the company to

explain itself better—and other things might open up other interesting possibilities (requests for less familiar Shakespeare plays or for work by other
Renaissance playwrights, for staged readings, for other educational sessions,
or for less "dramaturgical" concerns such as cheaper tickets for the elderly, or
for students, and so on). The dramaturg can learn valuable information about
where the audience is intellectually and culturally, all of which, if nothing
else, will help his sense of who is likely to see the next show, and will thus
inform all his future dramaturgical work from script editing to pitching the
next set of lectures.

21

The Dramaturg as Advocate

The end of chapter 20 serves as a segue into a more general observation on ways in which the dramaturg can serve as an advocate for various causes at various stages in the process and to a broad array of people. As the dramaturg's educational work with audiences is aimed, among other things, beyond the present production, so his advocacy of certain principles or ideas within the company can prepare the ground for future work.

One area in which the dramaturg can function as a species of advocate concerns contemporary issues to which the theatre can lend its voice as a tool for a general betterment of the culture at large. Mark Bly addresses this issue in terms of what he considers an American impulse to jettison the past before its lessons are learned: "I think dramaturgs are in a position to influence the kind of social, political and moral questions that are presented on our stages. But I fear that we too often abdicate our responsibility to push for productions that ask meaningful questions and challenge this tendency to 'forget'" (Moore 116).

In expressly theatrical matters, the two causes that I think the dramaturg should champion as much as possible are, on the one hand, the production of the most stimulating and innovative Shakespeare possible and, on the other, the continued and expanded role of the dramaturg itself. As will now be clear, though I am under no delusions about the power of the dramaturg, I think the two causes are related.

Dramaturgs need to speak up for themselves and their work, both to assert their own place in the company, and to advance the cause of dramaturgs elsewhere. But this is not simply about attaining more influence in the rehearsal room, getting more (or some) money for their labors, or gaining greater recognition for their contribution, though all those things are potentially valid reasons for standing up for the value of the job. It is also part of a much larger cause, one that asserts the value of an intellectual approach—or response—to life in general and art in particular.

It is a familiar joke between dramaturgs that they spend a good deal of their time explaining, both orally and in print, what it is that they do. This is, as I said at the beginning of this book, something that is inherent in the liminality of the position—its in-betweenness. But the shadowy nature of dramaturgy is also fostered by a tendency in theatre to mystify its own practices and to sidestep or dismiss outright the intellectual component in theatrical production, a mystification in which dramaturgs are sometimes made complicit in their very self-effacement. There is a popular myth that good theatre goes to an audience's heart, not their head, and that those who produce it (actors, directors, designers, and the like) use the former to the exclusion of the latter. This is a fiction, and one that encourages the cultures' larger skepticism about all things intellectual, one that the dramaturg needs to engage at all levels of his work, whether he is dealing with a director whose ideas seem poorly thought out, or with an actor who wants to intuit his way through the show without giving any ground to logic or a consistent idea, or with an audience who want to dismiss any intellectual exploration of the play as overreading. The dramaturg wears many different hats in the course of getting the show in front of an audience, but what unites his various roles and functions is *thought*, and it is this to which he is finally dedicated, perhaps even more than to the production itself because a commitment to thought in general goes far beyond the finite reaches of the production's run. Advocacy for the place of the dramaturg is, in its purest form, advocacy for the intellectual life and for the place of creative and theatrical endeavor within that larger frame.

Part of this intellectual commitment means standing up for issues and ideas that are not always popular or, in box office terms, easily marketable. As most theatre practitioners are well aware, Shakespeare's cultural status is a double-edged sword. On the one hand, his works have resilience and a perennial appeal as theatre at least in part because of that broader cultural status, the ubiquity of his plays in education, in the media, in the very language we speak. The downside of all this cultural prominence is the dubious baggage that audiences (and, in all honesty, scholars and theatre practitioners as well) bring with them to the staging of his plays. Indeed, the very familiarity of the plays is in some senses at odds with the purpose of theatre. Shakespeare is not new, and companies have to think, fight, create, and experiment to make him so. Unfortunately, it is unavoidable that companies feel a tremendous pressure (economic if not actually political) to produce Shakespeare on stage, which is itself not just familiar but safe, smooth, non-confrontational, predictable. Part of the dramaturg's job, speaking as an advocate for the intellectual dimension of the plays, speaking as someone whose name is unlikely to appear in the press reviews, speaking as someone who is of necessity largely separate from the hard truths of box office takings, is to press for shows that are fresh and exciting, shows that push the envelope, which knock the dust of Shakespeare and make a stand for the power of theatre.

That such a voice will often be drowned out by anxieties about staying fiscally solvent is to be expected and accepted, but the voice needs to be out there anyway. Somebody has to make the case for racial cross-casting, or for a visually dynamic *Two Gentlemen of Verona*, which throws naturalism out of the window, or for doing *Coriolanus* instead of yet another *Macbeth*, or, for that matter, of doing Jonson or Middleton instead of Shakespeare once in a while. Somebody has to make a case for doing reverse gender shows, or all female shows, for extensively doubled shows, for shows with no set, for shows that keep the house lights on, for shows that somehow—*anyhow*—challenge the standard assumptions about what an evening at the Shakespearean theatre is supposed to be about. Most of these battles the dramaturg will lose, but it is vital that the alarums and excursions continue, that those who make the decisions get to hear something other than the constant plea for inoffensive and lucrative productions. Few of the people who finally make the decisions got involved in theatre in order to be merely inoffensive and rich, and the dramaturg—unsullied as he tends to be by more pragmatic concerns!—is there to remind them of that. After all, if the theatrical event has nothing new to offer, if it does not generate real thought and real passion in its audience, if it is not genuinely funny or sad or maddening, if it does not throw light on the world in all its joy and terror, it is not worth doing. Too much of the Shakespeare staged today does none of these things, relying instead on the vague sense of elite cultural nostalgia to carry the indulgent, somnolent audience from curtain speech to curtain call.

As ever, of course, the dramaturg must pick his moments as well as his battles. An assault on the blandness of a show is of little use during tech week when nothing can be done about it, but at some point in the run or shortly thereafter the dramaturg needs to sit down with whoever runs the company and offer his assessment of the show's strengths and weaknesses. This is not to presume to the power to save or damn a show. It is the same position taken by every audience member who feels moved to write to the manager or artistic director after a production. Most of these people write to complain about what they thought was wrong or offensive. They rarely write to complain about what they found innocuous or dull. Enter the dramaturg. This is not to presume too much authority or, worse, claim some kind of superiority. It is merely to live the dramaturg's educational and collaborative function, and to try to make Shakespeare on stage something like what it once may have been: new, startling, and provocative.

22

Evaluating and Sharing your Experience

One would think that the only people to whom the dramaturg does not have to explain his job are other dramaturgs. This is not true. Since the dramaturg's role tends to shift depending on the company, the director, the play, and so forth, each experience is different and therefore potentially instructive to other dramaturgs. I therefore recommend that dramaturgs find means to talk to each other, both as people working in very similar situations (on different shows for the same company—even the same director—perhaps) or in radically different ones. On the local level, conversation between dramaturgs can help establish a sense of a particular director's interests and boundaries, company protocols, even preshow lecture demographics. Nationally or internationally, dramaturgs can learn of different approaches or principles that they could bring to the attention of the company manager, using precedents based on respected theatres elsewhere in order to lobby for more involvement at crucial stages of the process. In each case there are ideas to be shared that can help negotiate difficult predicaments or provide points of comparison.

Not only do dramaturgs have no union, they have no clear sense of collective identity, partly because their role is generally assumed to be vague or under construction. Connection to other dramaturgs can help build a sense of community that will aid in the development of a clearer sense of purpose. It is much easier to take a difficult stand with a strong-minded director when you know that other dramaturgs are making similar stands all the time, and it is similarly easier to accept finally being overruled when you know from other dramaturgs that this, too, happens all the time. I don't mean to suggest that dramaturgs are somehow more in need of a support group than other theatre practitioners, but I do think that their generally marginal nature can leave them with few people in whom to confide when things are frustrating or when they feel that their impact on the show is minimal. Reading on the subject of dramaturgy can help, of course, but there is no substitute for hearing of other people's triumphs and tragedies directly, and many will be glad to share their experiences.

An obvious way to track down other dramaturgs locally is through websites and programs from area companies. There will be a contact person who can put you in touch with dramaturgs the company has worked with in the past if their names are not listed. Dramaturgs can be found through publications on the subject (many are linked to colleges and universities and are relatively easy to find through institutional homepages) or through centralized locations. One such example is the dramaturgy Northwest website (http://www.ups.edu/professionalorgs/dramaturgy/), which is associated with the LMDA (Literary Managers and Dramaturgs of the Americas), the official website of which (www.lmda.org) contains useful resources and details of its own annual conference.

More formal ways to share one's experiences include lectures, and presence at conferences and publications ranging from chatty newspaper interviews to scholarly articles, all of which help the dramaturg to think about his experience, to codify it, and to render it useful both for himself in the future and for other dramaturgs. Writing about what we do helps to maintain an intellectual commitment to the work in its more self-reflexive form and (again through collaboration) furthers the general dramaturgical cause so that one day, perhaps, we will not need to constantly explain what it is that we do.

A necessary caveat to all this enthusiastic discussion and publication stems from my earlier observation of the way theatre practitioners sometimes mystify the process by which a show is made. Actors and directors do not always look kindly on having a dramaturg share their rehearsal room secrets with the world, particularly when they feel they don't come off well (or consistently well, or brilliant, or something: it isn't always clear in what ways a seemingly innocuous publication can give offense to someone mentioned within it). It's a bit like the indignation with which magicians respond to those tell-all books by renegade illusionists who show you exactly how the lady gets sawn in half, or how you make your tigers disappear. Some of this wariness is perfectly reasonable and some of it is rather less so, and since it is difficult to predict how a publication will be received by those who appear in it, it is probably wise to show them the piece before it goes to press. Their responses may be helpful; they may prompt changes, corrections, or rethinkings. Conversely, they may provoke indignation that coalesces into battle lines. In such a situation, the dramaturg then publishes or doesn't publish, with a better sense of the consequences for his personal and professional life. It goes without saying that a dramaturg needs to be trusted by those he works with and that a decision to print is sometimes the burning of a bridge. The situation, and the nature or justice of the objections to the publication, determines whether it is more important to maintain that bridge or to torch it in the name of personal integrity and a pursuit of truth.

When I set out to write this book I thought that I was merely describing a function, a process, or a set of related functions and processes collectively titled "dramaturgy." While I have deliberately attempted to maintain that sense of the descriptive, the book has wound up being more argumentative, more ideologically inflected than I had originally intended; not, I think, because I was caught up in the impulse to be polemic, but because I found that what I had considered merely descriptive actually *was* polemic. The more

I discerned my own agendas (about the self-authenticating nature of theatre, about editing principles, about the "Shakespeare police" stigma, and so on) the more I realized that the book enacts the dramaturgical collaborative function not by taking a middle ground so much as by taking a position that is consciously part of a discursive field: one that expects to produce debate and disagreement in a larger community. Though the most immediate part of that community is the subfield of Shakespearean dramaturgs, I see it as being considerably broader, encompassing scholars, theatre critics, actors, directors, designers, producers, managers and, most importantly perhaps, audience members. As such I hope this book—like the role it describes—will serve as a way to link these people and positions that are all too often deemed separate, belonging to different, even hostile, cultures and languages, that link helping to further the cause of staged Shakespeare as an energizing and connective cultural event. Dramaturgy, it seems to me now, has to use its intellectual authority—one grounded both in literary and performance theory as well as in the study of text and history—to seize the moral high ground so often claimed by "purists," opening the plays to a more truly performance-oriented spirit and a sociopolitical inclusivity so often denied when theatre is forced to do Shakespeare "by the book."

Notes

Introduction   The Shakespeare Police

1. Anthony Dawson's "The Impasse over the Stage," though a piece he has subsequently moved away from, accurately represents the sense of running into precisely this problem despite a spirited effort to get past it.
2. Geoff Proehl, a shrewd and experienced dramaturg, says that dramaturgs are "to some extent nowhere and defined by who they are not: not one or the other, not playwright or director; not audience member or performer, neither pure discursivity nor absolute theatricality; somewhere on the edge, somewhere on the margins" (Jonas et al. 136).

Part I   Principles

1. However much I enjoy the Tavern's work, I find its claims to historicity and proximity to the author's intent unconvincing in ways that seem significant for their unwillingness to employ a dramaturg. After all, a company that assumes the essential transparency and authority of the Folio as a performance text providing clues to the re-creation of a crucially authorial Shakespeare, and uses what they claim to be Renaissance theatrical technologies, leaves little for a dramaturg—as I conceive the role—to do, and less that would be logically consistent with their overarching production philosophy. While the Tavern does have staff who do double-duty as dramaturgs, the model of dramaturgy pursued is one that is supposedly geared to the recovery of a Renaissance original—a version of the Shakespeare Police that I consider to be at odds with good dramaturgy and, by extension, good theatre.
2. I'm quoting Marvin McAllister, who is currently The Shakespeare Theatre's literary manager. This, like all the company specifics cited in these pages, is taken from electronic correspondence with the respective organizations, which I initiated and compiled in 2004.

Section I   The Shakespearean Dramaturg: A Job Description

1. In fact, the late-eighteenth-century German intellectual Gotthold Ephraim Lessing provides a kind of starting point for much of what we today recognize as modern dramaturgy, though it was not until much later that Lessing's German legacy had any real impact on British and American theatre.
2. Manchester University Press, Fairleigh Dickinson University Press, and Arden all carry lines—though with different emphases—detailing the performance history of Shakespeare's plays. The quarterly journal *Shakespeare Bulletin* focuses exclusively on performance issues, and other major journals and yearbooks such as *Shakespeare Quarterly* and *Shakespeare Survey* devote space to performance reviews.

CHAPTER 1   THEATRICAL COLLABORATION AND THE CONSTRUCTION OF MEANING

1. For consideration of theatrical semiotics (the study of systems of signification: how meaning is constructed through the interpretation of signs of all kinds), see the work of Keir Elam and that of Marco De Marinis.

CHAPTER 2   THE TEXT/PERFORMANCE RELATIONSHIP

1. Thomas Kyd's *The Spanish Tragedy*, for example, seems to have been modified in subsequent revivals, one set of additional scenes being penned by Ben Jonson.
2. See Weimann "Playing with a Difference."
3. Stephen Greenblatt makes this case in his introduction to the Norton Shakespeare.
4. For consideration of this position see Robert Weimann's article "Playing with a Difference."
5. The obvious exception to this is *Macbeth*, which is unusually, and some would say problematically, short, particularly since some parts of the play seem to be additions by Thomas Middleton. Scholars such as David Bevington ascribed the play's brevity either to its being a text based on a particular performance, or one that has been censored.
6. *Hamlet* 3.2.38. All quotations are taken from the Bevington edition.
7. For a discussion of why some scholars prefer the infinite possibilities of armchair reading, see Berger.

CHAPTER 4   "AS IT WAS ORIGINALLY DONE": THE LOGIC BEHIND HISTORICAL RECONSTRUCTION

1. Even good historical re-creations of Renaissance theatres have a problem with these numbers since they invariably break local fire code regulations.
2. See Gurr, *The Shakespearean Stage*, for a thorough description of the construction and dynamics of Renaissance theatres.
3. For an intriguing examination of how Renaissance critique of acting styles suggests little about actual conventions and more about who were considered the good and bad actors of the day, see Menzer.
4. See, e.g., Howard, Rutter, and Orgel.
5. This sense of the differentness of early modern subjectivity is, of course, the bedrock of much New Historicist criticism and, using Foucaultian notions of power, informs the landmark work of writers such as Catherine Belsey, Katharine Eisaman Maus, and Stephen Greenblatt.

CHAPTER 5   THE NATURE AND USE-VALUE OF HISTORY

1. See, e.g., Lawrence Stone's *The Crisis of the Aristocracy*.
2. For evidence and discussion of this, see Keith Thomas, *Religion and the Decline of Magic*.
3. Again, see Belsey, Maus, and Greenblatt.

CHAPTER 6   AMBIGUITY AND POLYVOCALITY
IN THE PLAYS

1. Yachnin, "The Jewish *King Lear.*"
2. One of the most powerful moments in a GSF production of *Julius Caesar* to which I was attached was the lynching of Cinna the Poet. Since Cinna was played by a young black actor we anticipated the moment being read in terms of local history. In fact, more recent events (the show took place immediately after September 11, 2001) made many of the audience see in the Cinna episode some of the random acts of violence against Arab Americans which occurred at that time.

CHAPTER 7   AUTHORSHIP, AUTHORITY, AND
AUTHORIZATION

1. It is partly because of Shakespeare's iconic status that it is often assumed that audiences will attend Shakespeare—even supposedly unfamiliar Shakespeare or not-terribly-good Shakespeare—but won't attend plays by his contemporaries. The Bard's name suggests that the theatrical event will have special value, but it is this same assumption that feeds audience anxiety about the reality or purity of the Shakespeare that they experience in the performance.

CHAPTER 8   DIFFERENT LANGUAGES

1. For exceptionally revelatory examples, see works such as those by Dollimore and Halpern for expressly political studies of the plays in their original social and economic contexts. For studies of gender in the period and plays, see, e.g., Amussen, Adleman, and Kahn.

CHAPTER 9   WHY STAGE SHAKESPEARE?

1. To take a fairly obvious example, a production of *Twelfth Night* might interestingly explore issues of sexual orientation for audiences who would not go to see a modern show with overtly gay subject matter.

CHAPTER 10   PRELIMINARIES, CASTING, AND
DIRECTORIAL VISION

1. The LMDA assume dramaturgical presence at 50% of the rehearsals, and that the dramaturg will be paid approximately half of what the director is paid. Such a requirement is, obviously, aimed at professional companies with a significant investment in dramaturgical work. Most regional theatres and Shakespeare festivals pay lower rates, due both to more limited budgets and to a reduced expectation of the dramaturg's role.
2. See Dawson's *Hamlet* for a particularly good example.
3. Of course, whether the racism is perceived as inhering merely in some characters, as opposed to the play as a whole, the author, or his cultural context, continues to be argued by literary scholars. Whether the play is racist or is simply *about* racism then becomes a choice or series of choices made by the production, choices that will attempt to guide the audiences' readings of the subject matter.

4. For studies of cross-dressing in the Renaissance see Howard, Orgel, and Dawson ("Performance and Participation").

CHAPTER 11   THINKING ABOUT SCRIPT

1. See Homan's book *Directing Shakespeare*.
2. A complete works collection is available at http://www.the-tech.mit.edu/ Shakespeare.

CHAPTER 12   PREPARING THE SCRIPT

1. The first Quarto of *Hamlet* is, as I have said, much shorter than those of the second Quarto and Folio, and presents a very different play from the one with which most audiences are familiar. Whatever one thinks of the arguments about its proximity to an authorial or performed original, it at least raises some novel possibilities. For better or worse, the audience who comes expecting to hear the usual version of the "To be or not to be" speech will be surprised by the first Quarto's prosey "To be or not to be, ay there's the point,/To die, to sleep, is that all? Ay all: No, to sleep, to dream, ay marry there it goes,/For in that dream of death, when we awake,/And borne before an everlasting judge,/From whence no passenger ever returned,/The undiscovered country, at whose sight/The happy smile, and the accursed damned." This extract is reprinted in the textual notes to the Bedford case studies edition (Wofford 165).
2. Productions of *King Lear*, e.g., must choose between Edgar and Albany as to who has the last word in the play, choices based respectively on F and Q1, which have significant impact on the play's sense of closure.
3. Where knives and daggers replace swords in modern stagings, a word like "blade" often makes a good (and largely unnoticed) substitute for "sword." Firearms present different problems, of course, though one solution is to rely on Renaissance words like "pistol" or the less familiar "dag." "Gun" can work, though it sometimes strikes the audience as too modern (though it is, of course, used in the Renaissance, usually in reference to larger ordinance). A less specific term like "weapon" will sometimes do nicely.
4. Patricia Lennox, reviewing The Public Theatre at the Delacorte: Shakespeare in Central Park at the Delacorte Theater, New York, New York. June 24–August 15, 2003. Directed by Mark Wing-Davey.
5. The Pop Up Dramaturg joke came out of first night show parodies by the production staff and consists of someone with a droning voice and British accent emerging from a stage trap during the show to explain the significance or humor behind what was just said! Thus far, I (with my British accent) have managed not to take it personally . . .
6. The lines about the equivocator are also the only ones alluding to a crime (treason) whose relevance to the main plot of the play is insistently relevant.
7. Recent Georgia school board decisions to have high school students read Shakespeare only in modern translations have drawn tremendous fire in ways illustrating the point. Of course, reading a text (with recourse to footnotes and other sources of clarification) is an entirely different experience from the "real time" experience of hearing the words streaming from an actor's mouth. On stage, the justification for clarification through modernization is thus much greater, though the stigma of inauthenticity should give the dramaturg pause before he

doctors Jaques's "All the world's a stage" or similar "anthologizeable" moments. I don't consider such concerns to be pandering so much as acknowledging a reality about what Shakespeare is perceived to be, something that is a factor in why the audience is there at all.

8. I shall return to the question of how to think about the relationship between meter and semantics in the third subsection of chapter 17.

9. Kaiser's method, tested over several years at the Oregon Shakespeare Festival, includes speech prefixes and stage directions in the word count.

Chapter 13   Script-Editing Examples

1. The whole exchange about her gait being straight as a hazel twig, contrary to the halting he had gleaned from what he had heard is also obscure, but is most likely to be solved by physical business. Sometimes in performance she develops a momentary limp as a result of their tussling, though in the recent all-female performance at the Globe, the moment drew attention to a small physical deformity in the actress playing Kate (Kathryn Hunter). The joke was cruel, but contained a frankness that complicated the dynamic between them.

2. If the script needs further pruning, these two lines might be removed.

3. If this lengthy and confusing digression were not deleted, I would change "factor's" to "agent's."

4. "Meaner" does not necessarily suggest a woman of servant status, of course, but the alteration at least avoids the contemporary sense of "mean" suggesting "vindictive." Alternatives might be "poorer," "common," "homeless," or "vagrant."

5. The excised lines lose Aegeon's stoic lack of interest in his own survival, but the cut moves the lengthy speech forward. Of course, this lengthy stoic (or simply sad-sack) performance might be considered interesting of itself, even a source of humor.

6. Since the word "ship" has been used literally in the scene, it seems potentially confusing to now use it figuratively. The more prosaic use of "mast" helps avoid confusing the audience who followed the loss of the ship twenty lines earlier.

7. These lines might be removed in the interest of clarity and brevity, though some of the emotional value of the scene (and the play's continual sense of verbal balance) would be lost. If the cut was made I would modify the speech so that it ran: "So that, in this unjust divorce **my wife**,/Was carried with more speed before the wind."

8. This change is simply to avoid the curious time lapse in the use of "had siezed."

9. The Folio text contradicts the reference forty-six lines earlier, which suggests that Aegeon's wife took care of the younger son, not the elder.

10. The altered lines here are wholly my own invention in both form and sense. The lines as written can be unpacked, but they are fairly obscure at first hearing. Moreover, they do not explain exactly what happened to the eldest son or how Aegeon came to be looking for him. It might be argued, of course, that there is value in the notion that a lengthy speech of this kind leaves crucial details unexplained, but this is an overwhelming amount of information to throw at an audience right off the bat, and I think it should at least clarify the play's backstory.

11. I have a particular aversion to the Renaissance ejaculation of impatience "when!" To my ear it always sounds like something out of Monty Python or, at very least, suggests a sarcasm the original audience probably didn't hear.

12. The original is "adder" and refers to Britain's only poisonous snake, which is also called a viper, a term more familiar to American audiences.

13. The original says "that." "King" renders the line less ambiguous, and though it disrupts Brutus's evasion of the word, I think the resultant clarity worth the loss.

14. The original says "And." "Though" clarifies the logical movement.

15. The original says "affections." The closer translation is "passions," the emotional impulses that the Roman Stoics so mistrusted in that they unsettled the leader's basis in duty and reason, but "desires" leads more directly into Brutus's concern with Caesar's ambition.

16. This rather prosaically translates the original line "Will bear no colour for the thing he is." Such translation is, perhaps, not necessary, but I would at very least have it available to the actor as a simple explication of the original.

17. The folio says "first," a curious error, which suggests that Brutus's sense of the date is off by two weeks. I follow Bevington and others in making the change to "Ides."

18. The folio says "fifteen," which makes it the Ides of March already.

19. The original says "genius," a word certain to be misinterpreted by the modern audience.

20. The original says "the mortal instruments."

21. Cassius was Brutus's brother-in-law, a detail that does not get developed in the play, making this appellation (particularly since it is simply "brother") potentially confusing, if only for a moment. I don't want the tension of the encounter disrupted by the audience consulting with each other as to the relationship between the two men. Dropping the phrase disrupts the meter, but the nature of the subsequent staccato conversation makes the preservation of the blank verse of minimal importance.

22. The line is redundant and depends on an archaic usage of "favor" meaning "appearance." In fact, however, our production used the line as written because the dialect of the setting (Louisiana) and of the place of production (Georgia) both continue to use "favor" in this older sense.

23. These last three lines of the speech seem to me most difficult, particularly since they contain the Classicist reference to a realm of the underworld, but the apostrophic nature of the previous lines might be considered overly rhetorical for the privacy of the moment, as it is likely to be rendered in a modern production. A production may thus want to cut the speech after "They are the faction," which, left to hang by itself, might be nicely—albeit tacitly—laden with conflicted feeling.

24. There are many reasons not to cut these lines, but apart from suggesting that Brutus and Cassius are keeping private council even within the conspiracy, their consequences for the larger story are small. Removing them gets rid of Casca's clumsy metaphor about his sword (a piece of rhetorical performance strangely at odds with what we have heard from him and about him prior to this moment) and allows the audience to see the effect of the conference on Brutus, rather than moving him into some secret up-stage huddle while the preliminary issues of the conspiracy are negotiated. The cut also saves the modern production negotiating the anachronistic sword, and allows us to play the scene as night, not dawn.

25. This cut has been made for brevity and clarity. The meter is momentarily disrupted, but not jarringly, and it picks up perfectly in the next line.

26. The original word "palter" means, quite differently, "deceive" or "use trickery." The reworded line is thus more about resolution than straightforwardness, but it works adequately.

27. Brutus's point has surely now been made—and clearer made—by now, and while there may be mileage in the rhetorical excess, it is not necessary, and cutting it avoids the obscure word "cautelous," and the equally obscure phrasing that follows. The disruption to the meter here is greater than in the earlier six-line cutting, but it is playable since the previous remaining line ends the lengthy rhetorical question, thus calling for a beat before the new directive "Do not stain . . ."

28. Original says "meet."

29. Original says "annoy."

30. Original says "And."

31. The statement works perfectly well without these digressive lines anticipating Mark Antony's suicide, a speculation that seems less relevant in a modern production—particularly one set in a more contemporary locale—where suicide is a less immediate (or honorable) course of action when faced with defeat by association.

32. Original says "fear." An alternative rewriting that better preserves the original meter might be "We have no fear of him; let him not die."

33. Original says "augurers," a less familiar word than the alternative, which is often specifically associated with this play.

34. The original reads "Lions with toils and Men with flatterers." The adjustment maintains the meter while introducing the verb from the excised passage of quasi-proverbial folklore.

35. Again, cutting this passage loses the sense of Brutus taking the lead from Cassius, but that point has already been made. If the production wanted to maintain this dynamic (in preparation for their later falling out) but still prune the section, one way would be to reproduce the tail end of the section thus:

METELLUS CIMBER
Caius Ligarius doth bear Caesar hard,
[CUT: Who rated him for speaking well of Pompey:]
I wonder none of you have thought of him.
BRUTUS
Now, good Metellus, go along to him:
[CUT: He loves me well, and I have given him reasons;]
Send him but hither, and I'll fashion him.
CASSIUS
The morning comes upon's: we'll leave you, Brutus.

36. In the interest of brevity, some or all of this speech could be cut, though it redomesticates the scene in preparation for Portia's entrance.

37. In Portia's later speech in this scene she explains that at least part of her "weak condition" stems from the "voluntary wound" she has given herself as a show of her strength and constancy. I have cut this later reference, finding its Stoicism too alien to a modern production not expressly invested in the Roman setting to convey what it originally may have. In fact, I doubt many of the Renaissance audience would have found this overtly historical detail as especially impressive or sympathetic as the Romans did. The question of Portia's sickness thus needs rethinking. It could, of course, go unexplained (merely an illness, unconnected to the main action of the play), or it could be tied through the performance to an anxiety, even a depression, concerning her husband's mood and actions, though such a reading is contrary to Portia's show of strength in the original. In our production

the director chose to make her visibly pregnant, a device designed to heighten the domesticity of the scene, to emphasize the intimacy of their relationship and to generally raise the stakes for both husband and wife. As such, Brutus's remarks about her "condition" (and pregnancy is one of the few instances in which we still use that word specifically about bodily health) could thus stand unaltered, though their sense was, of course, quite different from that of the original.

38. The original word is "ungently," a word she reuses five lines later. "Secretly" changes the emphasis, of course, but our modern production's emphasis was not on discourtesy (as "ungently" suggests) but on the disruption of a relationship we assumed (in accord with the modern world of the show) to be intimate until it was disrupted. This is, of course, contrary to the respectful distance of the original, which pleads for such intimacy based on Stoicism and family background, both of which seem foreign to a modern setting and a contemporary audience.

39. The original says "humor" and is thus—in the Elizabethan sense—more physiological than simply mental. "Mood" loses a syllable, but the openness of the vowel sound permits a natural extension of the word so that the lost syllable is invisible.

40. The original says "physical." "Then" has been added for metrical purposes.

41. The original says "suck up the humors." The earlier word "unbraced" might be replaced by "undressed" or even with "without a coat," depending on how the scene is costumed. Neither this nor the following line are metrically regular, so some fluctuation in the pattern of the adjusted line is not so much acceptable as desirable, the lines indicating a break in the formality of the utterance.

42. The point has already been made, and cutting the lines removes the necessity of unpacking words like "rheumy" and "unpurged."

43. Deleting the textual prompt for Portia to kneel is, again, in accord with the more modern setting of the play and our related sense of this relationship's greater intimacy. I have thus altered Brutus's subsequent "Kneel not," to "Ask not," a shift of focus from the physical action of kneeling, which we have cut (and its implications of a premodern hierarchy within the home), to a desire to spare her the knowledge of what he is doing (analogous to some readings of Macbeth's lines to his wife concerning the murder of Banquo: "Be innocent of the knowledge, dearest chuck, till thou applaud the deed" [3.2.48–9].) Again, it changes the sense of the original but works logically in the context of what follows.

44. See previous note.

45. Though I am loath to lose the very playable anaphora these lines contain, the apparent assumption of female weakness is at odds with the modernity of our production, as is the patriarchal sense that strength is derived from her connection to the prominent males in her personal history. The change here is not merely political correctness, but the addressing of a historical condition in the same way the dramaturg might address any archaism that creates confusion, unease, or irritation that is not constructive. The issue here is not merely one of clarity but that antiquated attitude to female worth, and the ensuing "proof" of her ability to bear hardship through the voluntary wound seemed to us jarringly foreign and therefore distracting for an audience. During the performance of this scene I wanted to foreground the relationship between Brutus and Portia in the context of the conspiracy, not to have the audience worrying or silently arguing with the way Portia—or Shakespeare—rationalizes her value. While some critics will certainly take issue with such a change I think that for this production the sense of domestic harmony undamaged by the historical patriarchy of either

Rome or the English Renaissance is of more value. There is a real sense that this play and this production are both "about" issues of maleness and that this issue is underscored by the extent to which the female presence in the play is rendered marginal, but we felt that for this production there was real mileage to be gained out of Brutus being more sympathetic to his wife and to rendering the scene more clearly domestic and less Stoic.

46. See the note to Brutus's reference to Portia's "weak condition" on her entrance.

47. The original says "charactery," a word meaning "hand writing," something easily decipherable.

48. The decision to end the scene here is premised, again, on wanting to maintain the sense of domestic intimacy, partly to juxtapose this against that which begins the next scene between Caesar and his wife Calpurnia, so that the spousal relationships of the two men were pointedly aligned. The Ligarius episode, moreover, contributes little we do not already know. I do not believe that Brutus's dismissal of his wife in the original text on Ligarius's arrival necessarily reveals his previous promises to her to have been disingenuous, suggesting instead that the content of this next interview will be revealed to her hereafter, so I don't think that concluding the scene with the couple together significantly alters the original. Furthermore, strengthening the bond between them here creates the possibility of getting more mileage out of the news of her death later, and Brutus's show of Stoicism when he pretends to be receiving the news for the first time in front of his soldiers.

49. This is the only instance in which the text I have presented has no basis in an actual production.

50. Bell Shakespeare's 1998 production in Australia, directed by Barrie Kosky, gave the Fool a grab-bag collection of twentieth-century pop songs that had great ironic poignancy.

51. The original line ("Three merry men be we") is tonally quite different, but this anticipates the assertions of rank that follow it and underscores the play's associations with the Epiphany (January 6, or *Twelfth Night*). If the production was not being set around Christmas (and there is little textually internal reasoning for doing so) a different tack might be taken, one perhaps more clearly mirroring the sentiment of the original. I have a soft spot for Duran Duran's "Wild boys," Kool and the Gang's "Celebration," "Who let the dogs out?," Pink's "Get the party started," The Beach Boys' "I get around," or, just for devilment, Cyndi Lauper's "Girls just want to have fun," all of which have a suitably silly feel. I have deleted Toby's arcane jokes leading into the song.

52. The original line ("There dwelt a man in Babylon, lady, lady!") picks up drunkenly on Sir Toby's use of "lady" in the previous line, something the Tom Jones song does too. An alternative might be The Beatles' "Lady Madonna." Though the tone of the two songs is quite different, both evoke a certain distanced narrative and both are romantic, albeit in different ways. The high seriousness of the original is being mocked by its appearance in a scene of low comedy, something the use of the Beatles song might be seen as reversing (as the fairly modern popular culture collides playfully with the high art of the Bard). Both songs might be seen as playfully alluding as much to Olivia as to Maria. An alternative, and one that could be tailored to suit Maria's costume, might be Chris De Burgh's "Lady in red." Another might be Dylan's "Lay, lady, lay."

53. The original says "On the Twelfth Day of December," which may have been a Christmas song or a contemporary ballad.

54. The original word is "coziers."
55. The original is "Sneck up." A blunter and more contemporary alternative might be simply "Drop dead!" Less literally, "up yours" would work nicely.
56. The original says "an." An alternative might be "or."
57. The original word is "farewell." I have changed it to lead into the subsequent song.
58. Toby again uses the key word from the previous line to cue up his song, and though the number of popular songs featuring prominently the word "Farewell," has lessened considerably over the years (Bob Dylan's "Farewell Angelina," Billy Holliday's "Farewell to Storyville" and Alan Jackson's "Farewell party" are a few that come to mind). A similar effect can be created by replacing it with "Goodbye" in Malvolio's previous line, a word that easily generates all manner of suitably mawkish songs or others well-suited to parody: Elton John's "Goodbye yellow brick road," The Beatles "Hello, goodbye," The Backstreet Boys' "Friends never say goodbye," The Carpenters' "Goodbye to love," and The Jackson Five's "Never can say goodbye." With a little adjustment, the subsequent passage of dialogue between Toby and Feste could be made to fit whatever tune had been selected.
59. I have altered both the original text and the lines of the song so that something of the original script can be sung to the tune of the Cole Porter song.

Chapter 15 Tools of the Trade and Research Packets

1. Of course, factual information can produce difficulties, revealing inconsistencies or errors in the original material, as is the case in the way *Macbeth* conflates battles and compresses time. Actors and directors have a special ingenuity for finding ways to make sense of such problems, and are usually perfectly comfortable with acknowledging that some things don't always work as written. I think the advantage of making the places feel real to the company outweighs the difficulty of navigating occasional problems of fact or historicity.
2. Marvin Rosenberg's *Masks* books on the great tragedies are common and considerably better than most.

Chapter 16 Knowing Your Audience: Talking to Directors, Talking to Actors

1. Foucault's theory of the Panopticon centers on the idea of a prison building in which inmates feel that they are being continually monitored. Accordingly they adjust their behavior, effectively internalizing a kind of self-policing restraint, and eventually suppressing not just the bad behavior but the thoughts that produce it. The result, for Foucault, is the unsettling sense of people whose very subjectivity is stripped by the state of its autonomy, an image of how one's most private thought processes are shaped by larger material and ideological pressures. The Renaissance analogy draws on the idea that conscience was largely a Reformation construct that allowed the laity to discern right from wrong without recourse to larger ecclesiastical structure and doctrine. This, in the hypothetical *Measure* I am discussing, is bracketed with the similarly Renaissance fusion of church and state, and with the resultant sense of a continual surveillance of the people by both God and the secular authorities in ways reminiscent of the Panopticon.

CHAPTER 17   THE DRAMATURG IN REHEARSAL: A TEMPORAL BREAKDOWN

1. These are included in the Bedford/St.Martin's Texts and Contexts edition, edited by Frances Dolan.

2. Some of the historical materials listed here can be found in good editions of the play, particularly William Carroll's Bedford edition.

3. That the show was able to do exactly that at least occasionally was suggested by a colleague's remark that Caesar's line "Danger knows full well/That Caesar is more dangerous than he" (2.2.44–5) had never made more sense than when barked by a redneck politician to his wife across the breakfast table!

4. Bohemia, infamously, had no seacoast, a fact well known to many of Shakespeare's contemporaries. For my purposes this use of a known country with a fictional geographical feature nicely suggested a world for the play that is not real in the strictest sense, a country that is based loosely in fact but invokes a sense of distortion and make-believe. The idea helped the actors deal with the play's straying in and out of what was plausible/probable.

5. This kind of information is of obvious value to actors but is often maddeningly difficult to find. The Crystals' Shakespeare's Words contains a section under "money" which details monetary equivalents in England and elsewhere, though it does not attempt to suggest any modern correlations which might help the cast get a handle on the sums in question. I work from the assumption that an artisan in London in 1600 made approximately six shillings a week, about fifteen (predecimal) pounds a year. To find a modern equivalent is tricky but if we make that roughly balance a current wage of $30,000 a year, then the fifty pounds a year approximates a current annual wage of $75,000. By this reckoning the courtesan's ring which is given to Antipholus in *The Comedy of Errors* (valued at forty ducats or about ten English pounds) is worth $15,000 and the much contested chain Antipholus buys from the goldsmith is, at 200 ducats, worth about $75,000. These numbers are, of course, highly speculative, premised as they are on trying to make clear connections between two societies which are quite different in structure and their notions of money, but they can help give actors a clear sense of the stakes in a given scene. Even if the dramaturg does not want to get into modern equivalents, internal equivalencies (say, that Antipholus' chain is valued at fifteen pounds which is about what a London artisan could expect to make in a year) can be most helpful.

6. For an excellent overview of rank and class in Renaissance England, see Wrightson.

7. See his *Shakespeare's Advice to the Players*.

8. Rodenburg is nuanced and specific in her study of how to use the power of the Shakespearean line, so much so that her *Speaking Shakespeare* book should probably be considered required reading by any aspiring Shakespearean actor. I do think, however, that she overly valorizes the text and sometimes insists on certain readings of scenic moments based not so much on evidence as on an ideology of performance. Her claim, for example, that Hal's first soliloquy in *Henry IV part I* is motivated by something that happens in the previous scene and that his words represent a new awareness about the life he has been leading seem to me both groundless and reductive of playable options (25–6). Her general point about character/actor "readiness" and a certain proximity between thought and articulation is certainly valid and useful but it can be overstated once it becomes part of a manifesto. As with all such studies, then, I recommend that the dramaturg scrutinize their assumptions, agenda and logic, while recognizing that problems with their intellectual coherence does not prevent them from supplying useful approaches and methodologies.

CHAPTER 18   WRITING FOR THE AUDIENCE (SYNOPSES AND PROGRAM NOTES)

1. While program notes cannot determine how favorably a reviewer will respond to a show, they can help to clarify the production's logic or agenda, often using historical or textual evidence as a way of bolstering the show's authority. To those who think such a strategy disingenuous, manipulative, or reliant on non-theatrical elements as a way of justifying the production, I would say that, in part, I agree, but that I think programs *are* part of the show, and that most companies can ill afford to miss the chance of prompting a reviewer to embrace a production, reviews often having a direct impact on box office. As elsewhere in this book, I advocate the marriage of theory, principle, and ideal with the material conditions of practical theatre, in which latter category, company economics loom large. Sometimes local reviewers need to be taught if they are to shelve some of their misconceptions. If a program can inform a reviewer of what the company is going for and why in ways that make the resultant review more sympathetic, so much the better.

2. In some instances, program essays (often written by largely absent dramaturgs) that insist on a certain reading of the show may come as a great surprise to the actors in it, and an unpleasant surprise if the audience then reads the show through this textual lens.

CHAPTER 19   PROGRAM ESSAY EXAMPLES

1. The production for which this piece was written had Joe Knezevich doubling as Posthumous and Cloten.

CHAPTER 20   TALKING TO THE AUDIENCE

1. Some companies make texts available to audience members (especially season-wide sustaining patrons) when they book tickets for a show that involves a preshow lecture.

Works Cited

Adleman, Janet. *Suffocating Mothers: Fantasies of Maternal Origin in Shakespeare's Plays, Hamlet to the Tempest*. New York: Routledge, Chapman & Hall, 1992.

Amussen, Susan Dwyer. *An Ordered Society: Gender and Class in Early Modern England*. New York: Blackwell, 1988.

Barish, Jonas. *The Antitheatrical Prejudice*. Berkeley: California University Press, 1981.

Barton, John. *Playing Shakespeare*. London: Methuen, 1984.

Bate, Jonathan and Russell Jackson. *The Oxford Illustrated History of Shakespeare on Stage*. Oxford: Oxford University Press, 2001.

Belsey, Catherine. *The Subject of Tragedy*. London/New York: Methuen, 1985.

Berger, Harry Jr. *Imaginary Auditions: Shakespeare on Stage and Page*. Berkeley: California University Press, 1989.

Bevington, David, ed. *The Complete Works of Shakespeare*. Fourth edition. New York: Harper Collins, 1992.

Bulman, James C., ed. *Shakespeare, Theory and Performance*. London: Routledge, 1996.

Cardullo, Bert. *What Is Dramaturgy?* New York: Peter Lang, 1995.

Carroll, William C., ed. *Macbeth. Texts and Contexts*. Boston: Bedford/St.Martin's, 1999.

Copelin, David. "Ten Dramaturgical Myths." *What Is Dramaturgy?* Ed. Bert Cardullo. New York: Peter Lang, 1995.

Crum, Jane Ann. "Toward a Dramaturgical Sensibility." *Dramaturgy in American Theater: A Source Book*. Ed. Susan Jonas, Geoffrey S. Proehl, and Michael Lipu. Fort Worth: Harcourt Brace, 1997.

Crystal, David, Ben Crystal, and Stanley Wells, eds. *Shakespeare's Words: A Glossary and Language Companion*. New York: Penguin, 2003.

Dawson, Anthony. "Performance and Participation: Desdemona, Foucault and the Actor's Body." *Shakespeare, Theory and Performance*. Ed. Bulman. London: Routledge, 1996.

———. *Hamlet*. Shakespeare in Performance. Manchester and New York: Manchester University Press, 1995.

———. "The Impasse over the Stage." *English Literary Renaissance* 21 (1991): 309–27.

De Marinis, Marco. *The Semiotics of Performance*. Bloomington: Indiana University Press, 1993.

Dessen, Alan. *Elizabethan Stage Conventions and Modern Interpreters*. Cambridge: Cambridge University Press, 1986.

Dickens, Charles. *A Christmas Carol*. London: Heinemann, 1915.

Dolan, Frances E., ed. *The Taming of the Shrew: Texts and Contexts*. Boston: Bedford/St. Martin's, 1996.

Dollimore, Jonathan. *Radical Tragedy: Religion, Ideology and Power in the Drama of Shakespeare and His Contemporaries*. Chicago: Chicago University Press, 1984.

Elam, Keir. *The Semiotics of Theatre and Drama*. London: Routledge, 1981.

Greenblatt, Stephen. *Shakespearean Negotiations, the Circulation of Social Energy in Renaissance England*. Berkeley: California University Press, 1988.

———. *Renaissance Self-Fashioning: From Moore to Shakespeare*. Chicago: Chicago University Press, 1980.

Gurr, Andrew. *The Shakespearean Stage 1574–1642*. Cambridge: Cambridge University Press, 1992.

———. *Playgoing in Shakespeare's London*. Cambridge/New York: Cambridge University Press, 1987.

Hall, Peter. *Shakespeare's Advice to the Players*. New York: Theatre Communications Group Inc., 2004.

Halpern, Richard. *The Poetics of Primitive Accumulation: English Renaissance Culture and the Genealogy of Capital*. Ithaca: Cornell University Press, 1991.

Heywood, Thomas. *The English Traveller*. London, 1633.

Homan, Sidney. *Directing Shakespeare: A Scholar on Stage*. Athens: Ohio University Press, 2004.

Howard, Jean E. "Crossdressing, the Theatre and Gender Struggle in Early Modern England." *Shakespeare Quarterly* 39 (1988): 414–40.

Jonas, Susan, Geoffrey S. Proehl, and Michael Lipu, eds. *Dramaturgy in American Theater: A Source Book*. Fort Worth: Harcourt Brace, 1997.

Jonson, Ben. *Five Plays*. Ed. G.A. Wilkes. Oxford: Oxford University Press, 1981.

Kahn, Coppelia. *Man's Estate: Masculine Identity in Shakespeare*. Berkeley: California University Press, 1981.

Kaiser, Scott. "How Long Do You Think It Will Run?" *Shakespeare Bulletin* 22.3 (2004): 47–8.

Katz, Leon. "The Compleat Dramaturg." *What Is Dramaturgy?* Ed. Bert Cardullo. New York: Peter Lang, 1995.

Kennedy, Dennis. *Looking at Shakespeare: A Visual History of Twentieth Century Performance*. Cambridge: Cambridge University Press, 2001.

Lennox, Patricia. Reviewing the Public Theatre at the Delacorte: Shakespeare in Central Park at the Delacorte Theater, New York, New York. June 24–August 15, 2003. Directed by Mark Wing-Davey. *Shakespeare Bulletin* 22.1 (Spring 2004) 69–72.

McDonald, Russ. *The Bedford Companion to Shakespeare: An Introduction with Documents*. Boston: Bedford/St. Martin's, 2001.

Maus, Katherine, Eisaman. *Inwardness and Theater in the English Renaissance*. Chicago: Chicago University Press, 1995.

Mazer, Cary. "Solanio's Coin: Excerpts from a Dramaturg's Journal." *Shakespeare Bulletin* 21.3 (2003): 7–46.

———. "Rebottling: Dramaturgs, Scholars, Old Plays and Modern Directors." *Dramaturgy in the American Theatre*. Ed. Jonas, Proehl, and Lipus, 292–307. Fort Worth: Harcourt Brace, 1997.

Menzer, Paul. "That Old Saw: Early Modern Acting and the Infinite Regress." *Shakespeare Bulletin* 22.2 (2004): 27–44.

Moore, David, Jr. "Dramaturgs in America: Two Interviews and Six Statements: Dramaturgy at the Guthrie: an Interview with Mark Bly." *What Is Dramaturgy?* Ed. Bert Cardullo. New York: Peter Lang, 1995.

Onions, C.T. *A Shakespeare Glossary.* Ed. Robert Eagleson. Oxford: Oxford University Press, 1986.

Orgel, Stephen. *Impersonations: The Performance of Gender in Shakespeare's England.* Cambridge: Cambridge University Press, 1996.

Ripley, John. *Coriolanus on Stage in England and America 1609–1994.* Madison: Associated University Presses, 1998.

Rodenburg, Patsy. *Speaking Shakespeare.* New York: Palgrave Macmillan, 2002.

Rosenberg, Marvin. *The Masks of Macbeth.* Delaware University Press, 1992.

Rutter, Carol Chillington. "Learning Thisbe's Part—or—What's Hecuba to Him?" *Shakespeare Bulletin* 22.3 (Fall 2004): 5–30.

Stone, Lawrence. *The Crisis of the Aristocracy 1558–1641.* Oxford: Clarendon, 1965.

Thomas, Keth. *Religion and the Decline of Magic: Studies in Popular Beliefs in Sixteenth and Seventeenth Century England.* London: Weidenfeld & Nicolson, 1971.

Tillyard, E.M.W. *The Elizabethan World Picture.* Vintage, 1959.

Weimann, Robert. "Playing with a Difference: Revisiting 'Pen' and 'Voice' in Shakespeare's Theater." *Shakespeare Quarterly* 50.4 (Winter 1999): 415–32.

———. *Shakespeare and the Popular Tradition in the Theatre: Studies in the Social Dimension of Dramatic Form and Function.* Baltimore: Johns Hopkins University Press, 1978.

Wofford, Susanne L. *Hamlet. Case Studies in Contemporary Criticism.* Boston: Bedford/St. Martin's, 1994.

Wrightson, Keith. *English Society 1580–1680.* Piscataway: Rutgers University Press, 1992.

Worthen, W.B. *Shakespeare and the Authority of Performance.* Cambridge: Cambridge University Press, 1997.

Yachnin, Paul. "The Jewish *King Lear*: Populuxe, Performance, and the Dimension of Literature." *Shakespeare Bulletin* 21.4 (2003): 5–18.

FURTHER READING

Bank, Rosemarie. "Shaping the Script: Commission Produces a Bibliography of Dramaturgy." *Theatre News* (January/February 1983): 124.

———. "Interpreters, Dramaturgs, and Process Critics: A New Configuration for American Theatre." *The 1980 Winners.* Ed. Roger Gross. UCTA, 1981. 11–16.

Barba, Eugenio and Nicola Savarese. "Dramaturgy." *A Dictionary of Theatre Anthropology: The Secret Art of the Performer.* Trans. Richard Fowler. New York: Routledge, 1991. 68–73.

Barthes, Roland. "From Work to Text." *Textual Strategies: Perspectives in Post-Structuralist Criticism.* Ed. Josu V. Harari. Ithaca: Cornell, 1979. 73–81.

Beckerman, Bernard. *Dynamics of Drama.* New York: Alfred Knopf, 1970.

Bennetts, Leslie. "Stage Conference Asks What Is a Dramaturg?" *New York Times* June 23, 1983, sec. C: 15.

Bharucha, Rustom, Janice Paran, Laurence Shyer, and Joel Schechter. "Directors, Dramaturgs, and War in Poland: An Interview with Jan Kott." *Theatre* 14.2 (1983): 27–31.

Booth, Susan V. "Dramaturg in Search of an Axis." *American Theatre* (September 1990): 62–63.

Cardullo, Bert. *What Is Dramaturgy?* New York: Peter Lang, 1995.

Carlson, Marvin. "Theatrical Performance: Illustration, Translation, Fulfillment, or Supplement?" *Theatre Journal* 37 (1985): 5–11.

Castagno, Paul C. "Informing the New Dramaturgy: Critical Theory to Creative Process." *Theatre Topics* 3.1 (1993): 1–6.

Clay, James H. and Daniel Krempel. "How Does a Play Mean?" and "The Process of Interpretation." *The Theatrical Image.* New York: McGraw-Hill, 1967.

Davis, Ken and William Hutchings. "Playing a New Role: The English Professor as Dramaturg." *College English* 46 (1984): 560–9.

De Marinis, Marco. "Dramaturgy of the Spectator." *The Drama Review* 31.2 (1987): 100–14.

Ellwood, William R. "Preliminary Notes on the German Dramaturg and American Theater." *Modern Drama* 13.3 (1970): 254–8.

Hay, Peter. "American Dramaturgy: A Critical Reappraisal." *Performing Arts Journal* 7.3 (1983): 7–24.

———. "Dramaturgy: Requiem for an Unborn Profession." *Canadian Theatre Review* 8 (1975): 43–6.

Helbo, Andr et al. "Pedagogics of Theatre: Analysis of the Text; Analysis of the Performance." *Approaching Theatre.* Bloomington: Indiana University Press, 1991. 135–64.

Hornby, Richard. *Script into Performance: A Structuralist View of Play Production.* Austin: University of Texas, 1977.

Kott, Jan. "The Dramaturg." *New Theatre Quarterly* 6.21 (1990): 3–4.

LMDA Review. Newsletter Published by Literary Managers and Dramaturgs of the Americas.

McKenna, Maryn. "The Dramaturg: Towards a Job Description." *Dramatics* (April 1987): 28–31.

Pavis, Patrice. *Languages of the Stage: Essays in Semiology of Theatre.* New York: Performing Arts Journal, 1982. 27–8, 98, 100.

Schechner, Richard. "Drama, Script, Theater, and Performance." *Performance Theory.* Rev. ed. New York: Routledge, 1988. 68–105.

Schechter, Joel. "Heiner Muller and Other East German Dramaturgs." *yale/theatre* 8.2 and 3 (1977): 152–4.

———. "American Dramaturgs." *The Drama Review* 20.2 (1976): 88–92.

———. "Lessing, Jugglers, and Dramaturgs." *yale/theatre* 7.1 (1975/1976): 94–103.

Scolnicov, Hannah and Peter Holland. *The Play Out of Context.* Cambridge: Cambridge University Press, 1989.

Shyer, Laurence. "Writers, Dramaturgs, and Texts." *Robert Wilson and His Collaborators.* New York: Theatre Communications Group, 1989. 87–152.

———. "Playreaders, Dramaturgs and Literary Managers: A Bibliography." *Theatre* 10.1 (1978): 60–1.

Slavic and East European Arts 4.1 (1986). Issue devoted to dramaturgy in Russia and Eastern bloc nations. Edited by E.J. Czerwinski and Nicholas Rzhevsky.

Theatre 10.1 (1978). Issue devoted to dramaturgy. Edited by Joel Schechter.

Theatreschrift nos. 5–6 (1994). Issue devoted to dramaturgy.

Theatre Topics 13.1 (2003). Issue devoted to dramaturgy.

Theatre Symposium: A Journal of the Southeastern Theatre Conference 3 (1995): "The Voice of the Dramaturg." Issue devoted to dramaturgy. Edited by Paul Castagno.

Tynan, Kenneth. "A Rehearsal Logbook." *The Sound of Two Hands Clapping.* London: Jonathan Cape, 1975. 119–26.

Veltrusky, Jiri. "Dramatic Text as a Component of Theatre." *Semiotics of Art: Prague School Contributions.* Ed. Ladislav Matejka and Irwin Titunik. Cambridge: MIT Press, 1976. 94–117.

Williams, Raymond. "Argument: Text and Performance." *Drama in Performance.* Rev. ed. New York: Basic Books, 1968. 172–91.

Wolff-Wilkinson, Lila. "Comments on Process: Production Dramaturgy as the Core of the Liberal Arts Theatre Program." *Theatre Topics* 3.1 (1993): 1–6.

INDEX

actors, 98
  Renaissance, 47–8
  talking to, 65–9, 160–3, 171–5, 176
adaptation, 43–5, 88–9, 95–6
Alabama Shakespeare Festival, 11
Alchemist, The, 99–100
All's Well That Ends Well, 21, 193
Antony and Cleopatra, 194
archaism, *see* script editing
As You Like It, 87, 194, 203
Atlanta Shakespeare Company,
  10–11
audiences, 8, 26–7
  and adaptation, 43–5
  awareness of, 98–9, 179
  as collaborators in the construction of
    meaning, 31–2, 40
  difference from Renaissance, 48–9
  talking to, 205–8
  writing for, 183–91
authenticity, 43–5, 59–64, 156
authorial intent, 60–1
authority
  the dramaturg's, 2, 156
  the production's, 59–64
  textual, 90; *see also* text

Barish, Jonas, 59
Barton, John, 177
Bevington, David, 93, 104, 158
Bly, Mark, 82, 178, 179, 209
Brecht, Bertolt, 16

Cardullo, Bert, 2–4
casting, 82–6
  cross, 85–6
  politically controversial, 84–6
Chicago Shakespeare Theatre, 12
collaboration, 31–3
Colorado Shakespeare Festival, 11

comedy, editing for, 101–2, 116–27;
  see also script editing
Comedy of Errors, The, 155–6, 193
  doubling in, 83
  script editing example, 127–34
contract, 77–8
Copelin, David, 66, 80
Coriolanus, 88, 156, 210
Crum, Jane Ann, 23, 79
Crystal, David and Ben, 158
Cymbeline, 93, 109, 169, 186
  doubling in, 83
  program essay, 193–6

designers, 24, 26, 31, 69
  dramaturg's dealings with, 20, 66–7,
    71, 86–7
Dessen, Alan, 83
Dickens, Charles, 98
directors
  relationship with dramaturgs, 5–6,
    78–9
  talking to, 66, 81–7, 164, 175–6
doubling, 82–3
dramaturg
  abandoning ideas, 173–4
  addressing audiences, 23–4
  as advocate, 209–11
  and collaboration, 32–3
  as community, 212–14
  conflict involving, 6, 24–5, 32, 77,
    159, 170
  errors, 162–3
  ideologies of, 11
  as intellectual presence, 4–6, 80–1
  introductory remarks, 165–9
  invisibility of, 26–7
  as luxury item, 13
  as managers, 15
  as marginal presence, 25–6

as observer, 177–8
rehearsal presence, 151–2, 160–4,
    165–80
requirements of being, 19–24
responsibilities of, 15–27, 76
and Shakespeare, 17
status of, 24–7
and text, 45
as theatre practitioner, 2–6, 23
use of, in various theatres, 11–13
see also research; script editing
dramaturgical axioms, 18–19
dress rehearsals, 178–80
Duchess of Malfi, The, see, Mazer, Cary

editing, *see* script editing

Folio (1623) as acting text, 35–9, 89
    Hamlet, 96–7
    King Lear, 93, 100, 104, 110
    Macbeth, 113
    Tempest, The, 97

gender
    casting and, 85–6
Georgia Shakespeare Festival (GSF),
    7, 9–10
Guthrie Theatre, The, 12

Hall, Peter, 177
Hamlet, 19, 21, 40–1, 100, 188,
    194–5
    adaptation and, 43–4
    cultural specificity of, 54
    original staging of, 47
    performance choices and, 40–1, 112
    script editing of, 96–7, 102–3
    textual history of, 36–9
Henry IV Part I, 102
Henry V, 109, 156, 195
    historical and contemporary
      references in, 104–6
    playable opacity and, 100
    program notes and, 187
Henry VI Part II, 156
Heywood, Thomas, 38
history, 51–4
    dramaturg's use of, 20, 54–5,
      171–2
    of productions, 21, 79–80

of Renaissance stage, 46–50
    research into, 153–9
Homan, Sidney, 89

Jonson, Ben, 37, 210
    playable opacity in, 99–100
Julius Caesar, 108, 167–8
    program essay, 200–2
    script editing example, 135–44

Kaiser, Scott, 113
Katz, Leon, 80–1
King Lear, 57, 194
    contemporary songs in, 145
    program synopses, 189–90
    Tate's, 88–9
    textual problems with, 19, 38,
      93, 100

letter of agreement, *see* contract
light, *see* designers
literary criticism, dramaturg's use of,
    21–2, 79
Literary Managers and Dramaturgs
    of the Americas (LMDA), 77,
    151, 213

Macbeth
    cultural specificity of, 107
    doubling in, 83
    program essays for, 196–200
    research and, 154, 156–8, 167
    script editing of, 102–3, 106–7, 113
    textual problems of, 38
Mazer, Cary
    on dramaturgy, 160, 164
    and *The Duchess of Malfi*, 67
McDonald, Russ, 80
Measure for Measure, 21, 38
    rehearsal, 161
Merchant of Venice, The, 164
    racial issues in, 21, 58, 85
Merry Wives of Windsor, The, 154, 195
meter, *see* verse and meter
Midsummer Night's Dream, A, 171
    doubling in, 83
    inconsistent speech prefixes in, 39
    program essay, 202–4
    reassigning lines, 110–12
    and script editing, 108–9, 111–12

money
  for dramaturgs, 77–8, 209
  historical equivalents, 172
Much Ado About Nothing, 100
music, *see* designers

Nashville Shakespeare Festival, 11
National Theatre (UK), The, 12
North Carolina Shakespeare Festival, 11

Onions, C.T., 158
opacity, *see* script editing
Oregon Shakespeare Festival, 11
Orlando-UCF Shakespeare Festival, 11
Othello, 36
  and race, 58, 84–5

plays
  ambiguity in, 56–8
  printing and editing, 35–9
  see also individual titles
postshow discussions, *see* talkbacks
preshow lectures, 205–6
printing, *see* plays; text
productions
  historically "authentic," 46–50
  previous, 21, 79–80
  see also adaptation
program notes, 183–91
  examples, 192–204
  see also synopses

quartos, 36–8
  as acting texts, 39
  Hamlet, 36–7, 38, 96–7
  King Lear, 38, 93, 100, 104, 110

race
  casting and, 85–6
read-through, 169–70
rehearsals
  length of, 81–2
  phases of, for dramaturg, 165–80
  protocol in, 161–4
research, 22, 153–9, 165–9
resources for use in rehearsal,
    158–9
Richard II, 202
Richard III, 22, 40
Rodenburg, Patsy, 177

Romeo and Juliet
  adaptation of, 44
  inconsistent speech prefixes in, 39
Royal Shakespeare Company, The, 12
running time, estimating, 113
run-throughs, 178–80

script editing, 88–94, 95–114
  altering words in, 103–4, 109
  contemporary reference in, 105, 145–7
  cutting, 106–7, 135–44
  examples, 115–47
  famous lines, 108
  maintaining verse form, 109–10, 114
  opacity and archaism in, 97–104
  punctuation in, 112–13
  "purist" notions of, 44, 88–91
  reassigning lines, 110–12
  stage directions, 112
  stages of, 113–14
Seattle Shakespeare Festival, 11
Shakespeare
  ambiguity in, 56–8
  attitude to print, 37
  language, 18, 20, 70, 90–2, 172–3
  theatrical uniqueness, 17, 70–2, 210
  see also script editing; text; verse
      and meter
Shakespeare and Company, 12
Shakespeare Police, The, 2, 5, 13, 26,
    76, 163–4, 166, 214
Shakespeare Theatre The,
    (Washington D.C.), 12
songs, editing of, 145–7
sound, *see* designers
Stratford Festival, The, 12
synopses, 188–90

table work, 170–5
talkbacks, 206–8
Taming of the Shrew, The
  past and present performance
    approaches, 67–9
  research and, 167
  script editing example, 116–27
technical rehearsals, *see* dress
    rehearsals
Tempest, The, 173–4
  and race, 58
  textual problems with, 97

text
  construction and history of, 35–9
  dramaturg's sense of, 19–20
  editions, 93–4
  relationship to performance, 2–4,
      39–42
  see also script editing
theatre
  as constructor of meaning, 4, 31–3
  as local phenomenon, 7
  relationship to text, 2–4, 39–42
  Renaissance, 46–50
  temporal immediacy of, 17
theatres (specific)
  Renaissance, 46–50
  use of dramaturgs, 11–13
  see also Atlanta Shakespeare Company;
      Georgia Shakespeare Festival
      (GSF)
theory
  dramaturg's use of, 20–1, 64

Titus Andronicus, 58
tragedy, editing for, 102–3; *see also*
      script editing
Troilus and Cressida, 193
Twelfth Night, 92, 114
  doubling in, 83
  editing songs in, 145–7
  program essay, 192–3
Two Gentlemen of Verona, The, 210

Utah Shakespeare Festival, 11

verse and meter, 109–10, 114,
      176–7
voice coach, serving as, 176–7

Winter's Tale, The, 21, 85, 168
  doubling in, 83
Worthen, W.B., 59–63, 71

Yachnin, Paul, 57